Looking Glass
Universe

Looking Glass Universe

THE EMERGING SCIENCE OF WHOLENESS

by
John Briggs and F. David Peat

Illustrations by Cindy Tavernise

A Touchstone Book
Published by SIMON & SCHUSTER, Inc.
NEW YORK

Copyright © 1984 by John Briggs and F. David Peat
All rights reserved
including the right of reproduction
in whole or in part in any form
First Touchstone Edition, 1986
Published by Simon & Schuster, Inc.
Simon & Schuster Building
Rockefeller Center
1230 Avenue of the Americas
New York, New York 10020

TOUCHSTONE and colophon are registered trademarks of
Simon & Schuster, Inc.

Designed by Howard Petlack
A Good Thing, Inc.

Manufactured in the United States of America

10 9 8 7 6 5 4 3

Library of Congress Cataloging in Publication Data

ISBN: 0-671-63215-9 Pbk.

Picture credits

After Sheldrake: pages 212, 214
American Institute of Physics: pages 44, 73, 83, 176
John Briggs: page 266
A. Winfree: page 165

"Now, if you'll only attend, Kitty, and not talk so much, I'll tell you all my ideas about Looking-glass House. First, there's the room you can see through the glass—that's just the same as our drawing-room, only the things go the other way. I can see all of it when I get upon a chair—all but the bit just behind the fireplace. Oh! I do so wish I could see that bit! . . .

". . . Let's pretend the glass has got all soft like gauze, so that we can get through . . ." She was up on the chimney-piece while she said this, though she hardly knew how she got there. And certainly the glass was beginning to melt away, just like a bright silvery mist. In another moment Alice was through the glass, and jumped lightly down into the Looking-glass room. . . .

. . . "I'm sure I don't know," the Lion growled out as he lay down again. "There was too much dust to see anything. What a time the Monster is cutting up that cake!"

Alice had seated herself on the bank of a little brook, with the great dish on her knees, and was sawing away diligently with the knife. "It's very provoking!" she said, in reply to the Lion (she was getting quite used to being called "the Monster"). "I've cut several slices already, but they always join on again!"

"You don't know how to manage Looking-glass cakes," the Unicorn remarked. "Hand it round first, and cut it afterwards."

This sounded nonsense, but Alice very obediently got up, and carried the dish round, and the cake divided itself into three pieces as she did so. "Now cut it up," said the Lion, as she returned to her place with the empty dish. . . .

—Lewis Carroll,
Through the Looking-Glass

A Note About Terminology

In the interests of readability, the words "hypothesis" and "theory" are used interchangeably to mean scientific conjecture. This is the way these words are used in many scientific contexts, though a few scientists prefer more rigid distinctions. Some theories are generally accepted by scientists—for example, the Darwinian theory of evolution and quantum theory. Other theories might more properly be called hypotheses and have yet to pass what mainstream scientists would consider convincing scientific tests. The difference between a theory generally accepted by the scientific community and one that would be considered more conjectural will, we think, be clear from context.

For purposes of clarity, the authors have conformed to the (still) conventional usage of masculine pronouns with collective or general nouns (i.e., "the scientist"), but they would like the reader to be aware that they are not happy with this necessity and eagerly await the invention of a neuter pronoun for use in such situations.

Acknowledgments

The authors would like to thank Cindy Tavernise for her drawings, particularly for their note of whimsy, something every serious looking glass universe needs. We thank Dr. Frank McCluskey for his many hours of patient criticism and Dr. Steven Daniel, Dr. Jeff Gruen, and Jeff Briggs for their thoughtful readings of the manuscript. We extend particular gratitude to Dr. David Shainberg for his continual guidance in matters spiritual as well as technical.

To David Bohm, Rupert Sheldrake, and Basil Hiley who gave so generously of their time and insights we can only say we hope they find this book contains a fair and accurate representation of their views. We also express our appreciation to Dr. Karl Pribram for supplying us with his papers and to the staff of the Center for Research in Management Science at Berkeley for their help on the work of the late Dr. Erich Jantsch.

Finally, this acknowledgement would not be complete without mentioning the invaluable assistance of Florence Falkow and her painstaking copy-editing staff at Simon and Schuster.

Contents

Strange Reports

. . . the things go the other way.

Science and its sister, technology, are full of surprises—so many surprises it's difficult to be surprised anymore. Black holes, genetic engineering, dust-sized computer chips—what next? We're ready for anything. The theories and artifacts of science have long since become firmly established on our landscape, spreading and changing like a city's skyline. We've all become inhabitants in this city. Around us new structures rise, redevelopment projects take place as discoveries come and go. We take it in, rather jaded by this fast-paced and dazzling environment.

But lately, faintly, there has been a rumbling of the ground, a change in the light: mysterious signs. Strange reports reach us from people who have been working beneath the ground, in the deepest structures of the city, that they may have uncovered something, stirred something, which could drastically change the city and all who inhabit it. We have called the theoreticians who bring us these reports scientists of the looking-glass. They have a deep surprise in store for us, they say—deep because it is a surprise at the very foundations of science.

As yet, however, our city's planners don't seem worried. They assure us our basic structures and conveyances are safe. "Hard" evidence for looking-glass science is scanty, and its proponents are hopelessly outnumbered.

Who are these rebel theorists? What kind of great change do they portend? We will answer this question by focusing on four theories covering the spectrum of science from physics to chemistry to biology to the study of the processes of the brain. The theories have been proposed by eminent professionals exhaustively trained in the traditions of their fields and respected by colleagues for competence, precision, and past contribution. Yet the universe they collectively describe is so different from the scientific landscape we have grown up in that the situation can perhaps be compared to the Renaissance when the first modern scientists and great theoreticians Copernicus, Galileo, and Newton broke out of the labyrinths of medieval theology. That revolution took hundreds of years to unfold. This one could take only decades. In its wake it could overturn many of the deepest assumptions of such foundation-block theories as quantum theory, relativity, and the theory of evolution. It could bring us a world as removed from our current modern science as our science is from the occult certainties of the Middle Ages—perhaps even more removed, since the sense of the universe the looking-glass scientists attack is pervasive and ancient.

But in its subtle and complex detail, the message these new theorists bring us is actually simple: the flowing, swirling universe is a mirror.

TERRA COGNITA

TAVERNISE

Expeditions to the Edge of the Mirror

Oh! I do so wish I could see <u>that</u> bit!...

1. Theories: Spectacles of Speculum Speculation

In 1962 a physicist turned historian published a remarkable analysis of science. Ostensibly it describes how scientific theories and assumptions about nature are changed by social and subjective factors rather than "objective" criteria, and how there is no progress in science, only changing perspectives. These conclusions are certainly strange, because they run counter to some of our most cherished beliefs about science. But even the people who admire *The Structure of Scientific Revolutions* and the other works of author Thomas Kuhn may not realize how bizarre his revelations really are. Though Kuhn focuses on the history of science, by implication he has profound things to tell us about the reality science purports to investigate. As we shall see, Kuhn's analysis implies that science has always been a looking-glass enterprise in a looking glass universe—we just didn't notice.

Before we can understand what this means, however, we'll need a whirlwind tour of the ideas of science Kuhn overturns. Particularly, we need to get acquainted with the various key answers that have been given to the question: What is a scientific theory and how does it relate to the reality it describes?

The way to the looking-glass lies through a long-hidden gap in these answers. To pass through this gap requires recognizing and shedding some traditional attitudes about science. These attitudes might seem as well-worn and inevitable as our skin. So leaving them behind might generate some discomfort.

INDUCTION: SCIENTIST AS OBSERVER

One of the earliest renderings of the image most of us have about scientific activity comes from Francis Bacon, who described science as "induction." Bacon was a seventeenth-century courtier, statesman, philosopher, and man of letters. A few historians have credited him with the authorship of

Shakespeare's plays. Though that credit is certainly undeserved, Bacon was without question a powerful articulator of the break that scientists (then called natural philosophers) were making with the medieval view of nature.

Bacon didn't believe in an appeal to authorities such as Aristotle or the Church, although in his writings he was prudent enough to pay lip service to "divine revelation." Drawing inspiration from the systematic investigations of nature by early modern scientists like Nicolaus Copernicus, Tycho Brahe, Johannes Kepler, Galileo Galilei, William Gilbert, and William Harvey, Bacon argued that the method of natural philosophy should not be based on deduction. Medieval philosophers started with preconceived general ideas like Aristotle's "self-evident propositions" and from them deduced explanations for particular observations. Bacon said reasoning should go the other way, making observation from concrete things the starting point. By correlating concrete observations, Bacon thought, an investigator could arrive at generalities about causes and truth. As a crude example, suppose each time you go out in the rain you catch a cold. From a small number of instances you may reason inductively that you always catch cold after getting wet. From such a generalization could be born the hypothesis that getting wet in the rain is the cause of the common cold. Rejecting the appeal to authority, Bacon accepted *experience*, not premeditated dogma, as the true guide to knowledge.

But Bacon's argument from a limited number of cases to generalizations has obvious dangers. Not everyone who gets rained on catches cold. The cases observed may be exceptions or very special cases. Bacon was aware of this defect in his method.

To overcome the defect, Bacon proposed that science should use a systematic approach he called induction to uncover the regularities and orders of nature—nature's laws. By first gathering data, formulating a limited hypothesis, and then using this new knowledge to gather more data, the investigator could proceed in a careful and orderly way to uncover natural laws. Bacon believed that through understanding causes, his method would lead human beings to possess greater and greater power, not only over nature but also over society.

Bacon's early formulation of the scientific method helped shape our image of the scientist as objective observer and reasoner—disclosing eternal laws and attaining the essential knowledge needed to manipulate matter. Later, working scientists would add the idea of "controlled experiments" to help them isolate their observations and make them more precise.

The sixteenth-century thinker René Descartes had provided a strong metaphysical basis for the process of scientific observation and theory-making which Bacon defined. Descartes believed the universe is composed of two classes of substances: *res cogitans* (that is, the observer) and *res extensa* (the thing in nature to be observed). Observer and observed are essentially separate (though Descartes assumed they were ultimately held together by God). Things in nature were seen as objects or events which obeyed specific laws. The laws are the rules of causalities, how objects interact with each other. It was the job of the thinking thing or scientist to be *objective* (that is, to take the measure of objects) and discover these laws. He would then reflect the laws of objects' causalities in mathematical formulae. In his famous *Discourse on Method*, Descartes placed the highest value on discovering theories cast in a form such that all laws could be deduced from only a few axioms, like Euclid's geometry.

This idea was later to have a great effect on Newton, who inductively correlated the observations of Copernicus, Brahe, Kepler, and others to construct a theory containing only three elementary laws or axioms and an assumption about gravity. From these Newton was able to describe the entire motion of the universe, including the movements of planets, formation of tides, paths of cannonballs, and a host of other phenomena. Newton's *Principia* became a cornerstone of modern science, because it showed how much could be achieved by a rational, objective approach to nature. Newton's success also appeared to confirm Descartes' belief that nature is mechanical.

Newton's achievement was so impressive it led one of his scientific heirs, the nineteenth-century mathematician Pierre de Laplace, to forecast a day when a single mathematical formula would be found from which everything in nature could be deduced. "To it," Laplace said, referring to the mighty formula, "nothing could be uncertain, both future and past would be present before its eyes."[27]* Many modern scien-

---

*Sources of quotations are identified in the numbered References and Further Readings at the back of the book.

tists still echo essentially this view of the goal and promise of scientific progress. Recently, for example, cosmologist Stephen Hawking declared his optimistic conviction that all the major problems in physics will be resolved by the end of this century.

Newton's achievement also impressed the British philosopher John Locke, who is said to have been the first to use the word "scientific" in the modern sense. Locke equated "scientifical" with "certainty" and "demonstration" in "physical things."

Locke and his fellow empiricists Bishop George Berkeley and David Hume emphasized that all knowledge comes from the observer's sensations. However, the empiricists, especially Hume, while affirming the separateness of the observer from what he observes, and while stressing the importance of observation, had some doubts about whether induction could lead to certainty. Hume said that because nature has been rational and orderly in the past is no reason to suppose this will produce absolute knowledge in the future.

Hume's observation could be said to mark a turning point, after which the pure and absolute lines of science that Bacon, Descartes, Newton, and Laplace had pursued became increasingly blurred.

FALSIFICATION: SCIENCE ABOVE THE SWAMP

Karl Raimund Popper, an Austrian-born philosopher, was educated in the 1920s and '30s during an era which saw the emergence and acceptance of two profoundly disorienting new theories, relativity and quantum theory (about which we'll have more to say in the next chapter). These theories, which included such ideas as an "uncertainty principle" and the relativity of space and time, had the effect of calling into question Descartes' and Newton's strictly mechanical view of nature. They also cast doubt over whether or to what extent a scientist is separate from what he observes and hence raised questions about the meaning of objectivity.

From another quarter, Popper encountered the ideas of the Vienna school of logical positivists, particularly Ernst Mach, who argued, as had Hume, that science begins with sensations and observations. He therefore insisted that scientific theories can only express relationships between sense experiences. Mach's approach left uncertain whether the sense impressions and the theories based on them actually correspond to any truths of an objective universe "out there." Such a question, Mach believed, was metaphysical and should be rejected as unscientific.

After pondering these ideas and developments, Popper came up with a radical revision in the picture of science and its theory-making. He was able to rescue the ideal of scientific objectivity and our traditional conception of science. But to do so he had to redefine the role of theory.

Popper found it necessary to break with the older views by acknowledging that the scientist as objective observer isn't entirely separate from the things he observes. He interacts with them, using theories not just as explanations but as probes to provoke nature into something which the scientist can recognize as an observation. "Nature," he said, "does not give an answer unless pressed for it."[39]

The scientist starts with a theory which solves a "problem" (such as, "How do we get colds?"). He then presses nature by submitting this theory ("We get colds because we get wet") to a "test." Popper turned the inductive view on its head. Observing ten thousand white swans, he said, doesn't justify the statement "All swans are white." Observing ten thousand and one swans comes no closer to proving this theory. In fact, Popper decided, science is not really in the business of proving theories but of *disproving* them. A single negative result ("I haven't been wet for weeks and I still got a cold," or one black swan) can demolish a scientific theory, while no amount of successful experiments can logically prove its truth. All the scientist can do is put a theory to a crucial test in which certain of its predictions may be falsified. This approach is what makes science unique. Real scientific theory, unlike the theories of "alleged" sciences like astrology or Marxism, is set forth in a way that spells out observations and predictions which can be tested experimentally. The theory then stands or falls by the results of such tests. If the prediction fails, the theory has been falsified and must retire from the field. Popper argued it is no good simply trying to repair the fallen scientific theory by outfitting it with an exception or adjusting it to explain away its failure. Once a crucial prediction has been falsified, the theory must be abandoned or completely rethought. If a theory passes its crucial test, it has not been proved, only "corroborated," and the process of testing must continue. Popper was rather tough on theories.

He decided there are four marks of a good theory. First, its conclusions don't contradict each other. Second, what it claims to show isn't already buried in its premises. Third, it's "better" than previous theories in the sense of making scientific progress. Finally, though it might contain some elements which are unobservable, the theory as a whole must be capable of being "corroborated" or falsified in the real world by tests that lead to concrete observations.

Popper had met a formidable challenge. Quantum theory raised the specter that events at the subatomic level of matter might not be governed by absolute or determinate laws of cause and effect but by indeterminism, laws of chance. Even in the highly rational formalism of mathematics the scientific air in the twentieth century was getting strange. In 1931, Kurt Gödel had convincingly shown that in the very kind of mathematics scientists use and Descartes admired— axiomatic systems—there will always be statements which are true and consistent but which cannot be derived from a fixed set of axioms. To the classically trained scientist or mathematician this was like saying that if you put a pair of rabbits in an isolated pen and let them breed, after several generations there could be some who were brothers and sisters to other rabbits in the pen but were no relation whatsoever to the original pair. There are some ponderers of Gödel's proof who believe it is one of many developments which spell the end of rational science.

Popper, however, was able to respond to these bewildering new developments by redefining science and its theory-making process. Nevertheless, in his writings, he betrays a tinge of nostalgia and regret for the time of Bacon, Newton, Descartes, when it was possible to believe that nature was a machine and that by observing her closely scientists would gradually learn how she ticked. That time was gone, Popper realized:

Objective science has thus nothing "absolute" about it. Science does not rest upon rock-bottom. The bold structure of its theories rises, as it were, above the swamp. It is like a building erected on piles. The piles are driven down from above into the swamp, but not down to any natural or "given" base.[39]

Science is still progress, but now knowledge advances by a constant train of revolutions in which one theoretical structure is blown over by a falsifying test so a new and "better," deeper one can be erected in its place. Popper clings to the classic scientific belief that nature's laws are regular, final, and unchanging, though he has to admit they're covered by a "swamp" so deep science may never progress to the bottom. He rescues the image of science as a strictly "objective" pursuit by shifting ground. For the inductionists, objectivity was possible because nature's laws were absolute and clear, and thus human reason, if it observed things clearly, could come to see them. For Popper, objectivity is possible because of science's willingness to test anew even the most accepted theory to see if it produces some predictable observations.

Popper's new image of science became enormously influential for working scientists. Even one of the looking-glass theorists, Rupert Sheldrake, lays emphasis on having made his theory of morphogenesis explicitly falsifiable by key experiments.

PARADIGMS: THE RABBIT'S-EYE VIEW

Into this settled picture strides the analysis of Thomas Kuhn. Kuhn looks at the massive image of objective and progressive science as a historian, not as a philosopher. His primary interest is the question: In past and present, what is it that scientists really do when they make theories and perform experiments?—not what "should" they do. In exploring with this question, Kuhn spots a weird and previously hidden gap in our image of science.

Kuhn realized that revolutionary changes overturning theories are in fact not the normal process of science as Popper claimed, nor do theories start small and grow more and more general as Bacon claimed, nor can they ever be axiomatic as Descartes and Newton claimed. Rather, for most scientists, major theories, or "paradigms," are like spectacles which they put on in order to solve "puzzles." Every now and then a "paradigm shift" occurs in which these spectacles get smashed and scientists put on new ones that turn everything upside down, sideways, and a different color. Once the paradigm shift takes place in any scientific field, a new generation of scientists is brought up wearing the new glasses and accepting the new vision as natural or "true." Through these glasses, scientists then see whole new sets of puzzles.

Scientists themselves often use the metaphor of perception when discussing the evolution of a theory or the solution to a difficult problem: "Let's look at this from a different angle," "It all depends on how you view the problem," "I've been looking at this theory in a new way." Psychologists have constructed visual paradoxes which illustrate how perception can switch from one angle or "gestalt" to another. In the accompanying drawing, the duck is seen as a single image but then, in a sudden, nonlogical leap, it disappears and one sees the rabbit. In science, Kuhn suggests, this kind of shift takes place on a grand scale.

The process of discovering and growing habituated to a way of seeing was well known in the history of art long before Kuhn brought attention to a similar process in science. Consider, for example, paintings of the human form. The Egyptians pictured the human body with the face and legs in profile but the torso and eye facing to the front. Artists in the Middle Ages presented the body in a flat, elongated, linear form. By the Renaissance, the figure was painted to give the illusion of solidity and dimensionality. Postimpressionists emphasized the surface of the skin with its coloration and the effects of light. Braque and Picasso portrayed the planes of the body. In each case, some aspect of the visual form was given focus—shape, outline, solidity, movement, color, texture. Painting didn't "progress" or improve from the time of

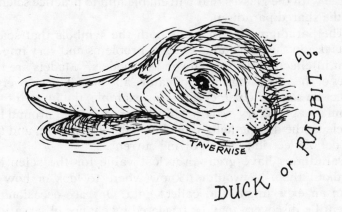

DUCK or RABBIT?

TAVERNISE

the Egyptians to that of Picasso; rather, the way of seeing—what art historian E. H. Gombrich calls the "schema"—changed. Each time a new schema is adopted, artists don a particular pair of spectacles and give attention to a particular visual aspect or way of seeing and portraying nature. At first the new schema seems unnatural and distorted to the public, but soon they become locked in and it begins to seem impossible to see things any other way.

The major element in the paradigm or spectacles a scientist puts on is the theory—quantum theory, relativity, the theory of evolution. But the paradigm also includes all the subtle presuppositions that surround the theory. Paradigms are acquired through a process like that which occurs when Johnny walks with his father in a zoological garden:

Johnny's education . . . proceeds as follows. Father points to a bird, saying, "Look, Johnny, there's a swan." A short time later Johnny himself points to a bird, saying, "Daddy, another swan." He has not yet, however, learned what swans are and must be corrected: "No, Johnny, that's a goose." Johnny's next identification of a swan proves to be correct, but his next "goose" is, in fact, a duck, and he is again set straight. After a few more such encounters, however, Johnny's ability to identify the waterfowl is as great as his father's.[24]

The scientific equivalents of Johnny's instruction are the textbooks and classroom activities which indoctrinate young scientists into their profession. In learning to solve the problems at the ends of the textbook chapters, a student gets the feel of what an acceptable experiment is in his field. He learns to see with the glasses that will enable him to practice science in the shared paradigm.

The paradigm also includes both the symbols that scientists use in their formulae to solve problems and very important intellectual devices knwon as "models." Models are images of the way things work in nature and often have powerful effect on scientific perception—for example, the image of the atom as a planetary system with electrons in orbit around the nucleus, the double-helix model of genetic structure, and the model of electricity as a flowing current.

Paradigms have great practical value for the scientist. Without them he wouldn't know where to look or how to plan an experiment and collect data. On rare occasions, a scientific discovery may be triggered by chance observation, but it is generally said that such observations are made only by the "prepared mind." In other words, the observations and experiments of science are made on the premise of theories and hypotheses. A medical student learns antomy so that he'll know how to see when he opens up the human body. If a lay person were to peer into an operation site, he would have little idea what he was looking at. Through careful training, the medical student learns how to "read" the human body, to recognize organs, to separate out the essential from the inessential.

Today most physicists "see" nature in terms of elementary particles. They believe reality is entirely constructed of such items as electrons, protons, neutrons, and neutrinos and that in turn these particles may be made out of more elementary entities called quarks. Such a hypothesis leads to construction of huge particle accelerators—machines specifically designed to observe the behavior and production of elementary particles. Without the hypothesis of such particles, it would be impossible to know how to design this equipment and what to do with it. But paradigm vision is double-edged. The elementary-particle paradigm which permits vision also restricts it. From the very nature of their design, the elementary-particle accelerators will only allow scientists to observe elementary particles. If some radically new phenomenon were present, it would not necessarily be "seen" with this device and might even be obscured by it.

A paradigm might also be likened to a map. In its early stages it is like one of those early-seventeenth-century maps of the Americas. Only the general shape is sketched in, riddled with minor errors and out of proportion, sea serpents and mythical dragons still lingering on the outskirts. The great mass of land area inside the outline is blank except for a few rivers and mountain ranges. But the overall sense of the continent is there. The task of "normal science" is to fill in the blank area and correct the discrepancies in the shape of the land, expelling the dragons and giving a progressively more detailed picture. The map supplies the general outlines and directions for this activity.

This was the state of affairs following establishment of the "classical" paradigm in physics. Much of the paradigm had come from Newton's *Principia*, which provided a general continental shape and a few significant landmarks. It remained for eighteenth- and nineteenth-century normal science to refine the picture. Some scientists, for example, worked to bring the divergent techniques and theories of the day into line with Newton; others, employing new inventions and improving old ones (such as the telescope), labored to make observations that would fill in the paradigm and give it concrete substance. Telescopic observations of orbits of planets differed from the theory, so still other scientists fine-tuned the theory to match with actual experiment.

In the process of shaping and correcting their paradigm, scientists make other discoveries. In setting out in a direction, perhaps with a specific destination, they are like those explorers who sought the northwest passage to the Pacific and found the Great Lakes along the way. The development of the mathematics of hydrodynamics (the motion of fluids) and solutions to the problem of vibrating strings were discoveries which took place within the general if vague dimensions of the map laid out by Newton's *Principia*.

By the end of the nineteenth century the map of the Newtonian paradigm looked almost as detailed as nineteenth-century maps of the Americas. It seemed only a few areas were left to explore.

But as they probed these areas, the puzzle-solving mapmakers of normal science began to experience disturbing difficulties. They saw light behaving like both waves and particles and electrons jumping from one orbit to another in the atom instantaneously. It was as if the explorers with their surveying instruments suddenly were falling through the ground they were mapping into another reality. The appearance of "anomalies" which could not be fitted into the classical Newtonian picture of the world was bringing on what Kuhn calls a "crisis" in the paradigm.

It was here, as he explored what happens around the periods of paradigm crises, that Kuhn discovered a serious if well-hidden breach in the traditional objective and progressive view of science.

He noticed that during times of crisis, new theories arise to explain anomalies. These theories vie with one another for the honor of becoming the new paradigm. Meanwhile, adherents of the old paradigm in crisis fight to retain it against the revolutionaries who are outrageously explaining anomalies by treating nature as if she were a rabbit or squirrel instead of what every self-respecting scientist knows she is: a duck.

According to Popper, such controversies should not exist. Scientists faced with two competing theories to explain anomalies would simply choose the "better" one—the one which could survive the toughest tests. Kuhn realized that, in practice, this is impossible. Paradigm shifts are such drastic alterations of perception that advocates of theories will not even agree on what constitutes a valid test. Such so-called "objective" criteria for choosing a theory as accuracy, consistency, scope, and simplicity all contain subjective factors. For example, even if scientists could agree that "simplicity" should be a standard for choosing between theories, they might well disagree on which theory is really simpler. The parties in paradigm debates speak different languages and "talk through each other." They wear such different glasses that even if they should read the same compass they wouldn't agree on the needle's direction.

Kuhn concluded that Popper's falsification standard of objectivity was a myth. Popper had charged astrology as unscientific because it refuses to accept any failure of prediction as disproof of astrological theory. But Kuhn saw that science often does the same thing. Meteorologists don't abandon their theories when the predicted sunshine doesn't arrive. They claim what the astrologer claims—too many variables and uncertainties in the system. In fact, most scientists hardly ever see an anomalous experimental result as a challenge to the paradigm that lies behind it.

Consider, for example, what happened in the case of the missing quarks. By the 1960s, a theory had been proposed which said that an elementary particle is composed of three more elementary particles called quarks. Each quark was said to have a fractional electrical charge. The properties of these quarks were predicted, and all over the world physicists set up experiments to detect them.

No quarks were ever detected. Did this mean that the theory had been falsified? Not quite, for the theory had only predicted the existence of quarks, not that anyone would actually see one. Moreover, as increasing numbers of elementary-particle experiments were performed, it became clear that three quarks could not explain the new results. Was the theory abandoned then? Had it been falsified? Not at all; it was simply expanded to include six quarks instead of three. In addition, it was now proposed that quarks are in principle unobservable.

Although the theory of falsifiability of scientific theories may be logically attractive, it has little to do with how scientists actually proceed. Real theories are difficult to falsify in practice because any experiment involves a range of possible errors and a complex set of processes and procedures—so there is always room for modification and adjustment of assumptions. In most cases, scientists would prefer to adjust a theory and thereby maintain its overall structure rather than simply abandon it because it doesn't fit certain facts. Often they would prefer to think that the failure of an experiment is their own failure to fit the paradigm properly. Why? It would seem that for a scientist (and perhaps for the rest of us) there is one thing *even* worse than personal failure: finding out that the reality you have been looking at and working in all these years is in fact some *other* reality.

If falsifiability isn't a standard assuring the objectivity of scientific theory, what is? Kuhn did discover scientists as a whole applying some standards for major theory acceptance. First, the new theory's explanation must answer a significant number of questions already answered by the old paradigm, though by no means all. Some of the questions which were previously resolved might become open questions in the new theory or might be considered irrelevant. Second, an acceptable theory has to provide enough puzzles to keep alive scientific research. A theory which resolved all questions wouldn't be of much interest to scientists since it would leave no puzzles and no research jobs.

These are the criteria Kuhn observed, but they hardly assure that the choice of theory will be objective. That's because Kuhn has, in fact, exposed an undreamed-of crack in the long-held conviction that science is leading us toward the ultimate truth about nature's laws. As we squeeze through this crack we may want to contemplate how strange this universe really is.

In the eighteenth century, John Dalton's new chemical theory of compounds featured a completely different set of assumptions from those which guided chemists of his day. When Dalton searched the chemical literature to find evidence for his theory, he found some examples of reactions that fitted his hypothesis but many that didn't. For instance, one experimentalist famous for the accuracy of his measurements had analyzed the oxygen/copper proportions in copper oxide as 1.47:1. Dalton's atomic-weight theory demanded a ratio of 2:1. Kuhn says of this:

It is hard to make nature fit a paradigm. That is why the puzzles of normal science are so challenging and also why measurements undertaken without a paradigm so seldom lead to any conclusions at all. Chemists could not, therefore, simply accept Dalton's theory on the evidence, for much of that was still negative. Instead, even after accepting the theory, they had to beat nature into line, a process which, in the event, took almost another generation. When it was done, even the percentage composition of well-known compounds was different. *The data themselves had changed.* [Emphasis added.][25]

Kuhn says, alluding to the paradox of perception, "What were ducks in the scientist's world before the [paradigm] revolution are rabbits afterwards. . . . What is more important, during revolutions scientists see new and different things when looking with familiar instruments in places they have looked before."[25]

Imagine again for a moment that we are at the initial stages of a paradigm crisis. The old paradigm can't account for certain anomalies, strange data, that are being reported. Two new theories appear which offer quite different explanations for the anomalies. These theories represent different realities, different sets of spectacles, the outlines of vastly different maps. After a while, one of these theories begins to gain support in the scientific community. The reasons for the support are not objective. Scientists like the elegance or simplicity or explanatory range of the theory. This backing leads to experiments, and soon evidence that "corroborates" the theory begins to roll in. The more evidence accumulates, the more support among scientists—particularly younger scientists—the theory gains. Soon reality begins to take on (or be seen from) the new direction. Like painters who have suddenly begun to see an aspect of the human form, scientists begin universally to see and test for certain features of reality and ignore or reject others.

But what if support had been given the other theory? Would evidence have flowed in for that? Could reality have turned in that direction? According to Kuhn, it is definitely possible.

This means, of course, that there is no real scientific "progress." A new paradigm doesn't build on the paradigm it replaces; it turns in an entirely new direction. As much knowledge is lost from the old paradigm as is gained by the new one. What scientists in the Middle Ages knew about the universe, quantum scientists no longer know. Instead we "know" a *different* universe.

The idea that there's no progress in science may be a difficult one to accept, because we look around at television and computers and trips to the moon. Surely these technologies are the signs of scientific "advance." This conviction is so infused in our image of science it seems mad to suggest otherwise. But technology is basically engineering and involves the invention of instruments and techniques. These play an important role in science by making new observations possible and providing models. While it's true that scientific paradigms often suggest directions for inventors to go to create new instruments, most engineers would admit that new technologies often come more out of a push and pull with reality than out of theories. And do new technologies imply that we really understand nature "better" now than we did in the fourteenth century? Or better than traditional cultures such as those of the American Indian? Are technologies really an advance? It might be argued that in many ways we are less in tune with nature than we ever were and that even our technological knowledge has not so much advanced us as produced new levels of ignorance. Some of this ignorance seems at least as treacherous as the forces of nature we believe we've controlled with our technologies. Those who want to make this point would not have to go far to find evidence of an increasingly confused understanding about nature manifesting itself in pollution, nuclear wastes and armament, urban congestion, and our destruction of planetary wildlife.*

If Kuhn's analysis is true, it also demolishes one of the primary underpinnings of the scientific method. The whole idea of a scientific experiment rests on the assumption that the

* The daring idea that technology is not a proof of progress can be illustrated by considering the American Indian. It is often said that medical technology has dramatically improved our general health, increased the average life span—and this seems undeniable proof of progress. It turns out, however, that when colonists came over from Europe they found the American natives were on average tall, healthy, and exceptionally long-lived. The Indians possessed a sophisticated knowledge of herbal medicine and psychological methods of disease cure which are just now being appreciated by modern American culture. European technology degraded rather than advanced the Indian culture and much of the Indian naturopathic knowledge has been lost. One might say that the idea of technological progress works if the frame of reference is sufficiently narrow. Western medicine today may be "better" than Western medicine one hundred years ago. But is it better than other kinds of medicine, other paradigms?

observer can be essentially separate from his experimental apparatus and that the apparatus (in Popper's term) "tests" the theory. Kuhn shows that the observer, his theory, and his apparatus are all essentially expressions of a point of view— and the results of the experimental test must be expressions of that point of view as well.

Kuhn's analysis of scientific history strips us of the traditional preconception that science is objective. Does that mean, then, that it's *subjective*? ("It's all in our minds"?) It can't be. Obviously observers don't just dream or control reality with their theories. After all, experiments fail, paradigms are eventually overturned. There must be something at work *other* than the fantasies of the human mind. But if science is neither objective nor subjective, what is it? What is the relationship of the scientist to this universe he observes? What is this universe?

Between the lines of Kuhn's analysis we push through the gap in the traditional view of science and down a narrow tunnel. We now stick our heads out into a fog-shrouded landscape—shimmering, infinitely subtle, and new. In this landscape we see scientists as they move from paradigm to paradigm like rabbits in a magic show, seeming to discover in their movement that *the very laws of nature are protean, changing with each new paradigm.* As the scientists shift paradigms, even the data change (remember Dalton). And as it unfolds, a paradigm seems to *generate* (not just uncover) anomalies which destroy it, leading to others. Thus here, through the streaming mist, we seem to glimpse the strange possibility that *the changeableness of nature's laws may be relative somehow to the activity of scientists' looking*. Observer and observed appear to influence one another, the scientist like a whirlpool trying to study the flow of water. Here we have left behind, with Bacon and Descartes and Popper, a universe where the observer *observes* the observed and have entered a looking-glass, a universe where, in some way (we can only see this part very dimly now) the *observer is the observed*. We may reflect that if this is so, then we may have discovered a universe that is whole.

But the fog closes over and we are left with only this brief and tantalizing glance.

Nevertheless, we know Kuhn's view has taken us far away from Locke's definition of science as "certainty" and "demonstration" in "physical things." The traditional views of science assumed scientific inquiry to be guided by the concrete objects and movements of nature, the physical as opposed to the metaphysical. Kuhn has shown us that in science the physical and metaphysical, facts and ideas, matter and consciousness, experimenter and experiment, are somehow one movement. To understand nature, he seems to imply, we will need to understand much more than we now do about what this movement is and how it works. For without such understanding it seems evident that the meaning of our scientific experiments will be hopelessly confused. Specifically we will need the answer to questions like: How are observer and observed, matter and consciousness, related? How is it nature's laws can be so stable for long periods of time (between paradigms), and on what principle do they change?

Kuhn's approach takes us far but cannot begin to answer such questions. That must remain for looking-glass science. And at this point we should remember that Kuhn's thesis is itself a paradigm and thus sees new things about science while being blind to others. It is important to us here because it constitutes a first, vague step into an unfolding universe in which the observer is the observed. It offers insight into the social forces of science which will undoubtedly decide whether any theories to explain an observer-is-observed universe will eventually be accepted. And it also forewarns us that to see this new universe we will have to put on spectacles likely to make the familiar as unfamiliar as our image in a carnival's wall of mirrors.

We turn now from science's history and philosophy to search for the looking-glass in those "hard" problems which working theoretical scientists grapple with as they try to understand the birth, death, and transformations of matter. We move to the scientists themselves and the signs of an uncanny universe their probings have encountered.

2. The Second Expedition

In the early decades of this century, two scientific expeditions, both mounted by physicists, reached the edge of the looking-glass. These explorations constituted the time of the great paradigm shift from Newtonian physics, and they were responsible for establishing the paradigms of quantum mechanics and relativity. Neither of the expeditions passed entirely through the glass, but in developing their theories both saw new sights and scouted trails which prepared the way. The first of the two expeditons to reach the edge was led almost single-handedly by Albert Einstein; the second by a collection of adventurers including Niels Bohr, Werner Heisenberg, and Erwin Schrödinger. We will open our inquiry into the "hard" evidence for the looking glass universe with this second expedition, because it produced the most vivid results. The findings of Einstein's earlier exploration can then be seen in its light.

THE SHAKEUP OF BOHR'S ATOM

It's impossible to say just when the quantum revolution started. By some calculations, it began at the turn of the century—but it took well over two decades before anyone really suspected that a paradigm shift was under way. We will start our own telling of the story about the time of this recognition with one of the undisputed expedition leaders, Niels Bohr.

The ancient Greeks believed the atom to be the ultimate unit of matter, the indivisible part out of which all other parts are made. Early-twentieth-century physicists had discovered that this indivisible atom itself had "parts," the proton and electron, and they devised various pictures of how these fit together.

The great experimentalist Ernest Rutherford proposed the most convincing model. Rutherford said the atom was like a tiny solar system, with a massive central core surrounded by lighter orbiting electrons.

Niels Bohr had thought about this appealing model and realized there was something seriously wrong with it. When the laws of classical physics were applied to the miniature solar system, a paradox appeared. Calculations said the orbiting electrons should give off energy and spiral down into the nucleus. Measured on human time scales, this orbital decay would happen in the blink of an eye. Theoretically, therefore,

EARLY MODEL OF THE ATOM

One of the early-twentieth-century ideas of the atom imagined protons and electrons held inside atomic structure like raisins in a plum pudding.

all atoms should collapse in a moment. And, since everything is composed of atoms, there should be no stability in all the universe. Such things as mountains and stones should be impossible. Obviously this wasn't the case.

Bohr therefore proposed a triumphant and far-reaching new model to resolve this paradox. He discovered it by combining two clues.

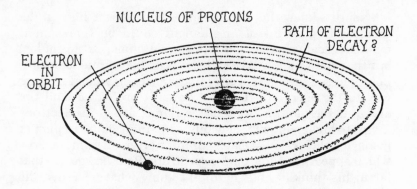

NUCLEUS OF PROTONS

PATH OF ELECTRON DECAY?

ELECTRON IN ORBIT

RUTHERFORD'S ATOM

In our solar system, the planets are continually losing energy and spiraling gradually toward the sun. We don't notice this because the scale of our solar system is so large. Bohr saw that Rutherford's atomic solar system, because its scale is so small, would seem to us to collapse in a moment.

The first clue was the spectrum of hydrogen. If a pinch of salt or other chemical is burned, it gives off light. When this light is passed through a kind of prism, it reveals a pattern of bright lines—the chemical's spectrum. Each chemical or element has an identifiable pattern to its lines—a "fingerprint." This type of spectrum is called an "emission" spectrum because it records the energy (i.e., light) given off by atoms when they're excited (i.e., burned). Scientists also work with an "absorption" spectrum which is formed when energy is taken in by atoms. The absorption spectrum consists of dark lines in a bright background and is a negative image of the emission spectrum.

The absorption spectrum of hydrogen. The hydrogen spectrum had received a great deal of attention in the late nineteenth century and the formulae for its fine lines were well understood, though what the lines corresponded to inside the atom remained a mystery until Bohr.

Scientists knew that the distances between the lines in the spectral fingerprint of an element could be represented mathematically by particularly simple numerical formulae, but they didn't know what the lines or the formulae meant. Bohr guessed that the spectral lines must correspond in some way to the energy of the electron as it orbited around the atom. As the atom is excited in a flame, one of its electrons jumps to higher and higher orbits, something akin to a pianist going up the scale. At each jump it gives off some energy which appears as a line on the spectrum. Bohr had gotten that far in his thinking, but he didn't know what governed the electron orbits. Why were they spaced according to such simple formulae? Why did the electron take only fixed (discrete) steps instead of spiraling up or down to its new orbit the way a planet or satellite does?

To answer these questions, Bohr turned to his second clue: the quantum of Max Planck.

At the turn of the twentieth century, Max Planck had every reason to expect he would be among the last of the world's theoretical physicists. The thinking at the time was that the Newtonian paradigm had solved most of the major problems in physics and as soon as the remaining, rather minor, ones were cleared up, there would be nothing much left for physicists to do. It was Planck's fate to change all that.

In the early nineteenth century, Thomas Young had demonstrated that light is a wave. By the end of the century, Scottish physicist James Clerk Maxwell had shown that light waves are electromagnetic ripples of energy and that electromagnetic waves include not only visible light but other energies ranging from low-frequency radio waves to high-frequency gamma waves. The theory worked well except when physicists began to calculate the total energy inside a heated "black box." Absurdly, the calculations showed the energy as infinite.

To solve this paradox (what turned out to be a paradigm anomaly), Planck proposed the bizarre idea that light energy can be emitted and absorbed in discrete or separate units he called "quanta." The problem with this was it contradicted Young's theory that light travels as continuous waves. Though Planck balked at the implication of his discovery, a young physicist, Albert Einstein, won a Nobel Prize for showing that indeed energy does possess a particlelike nature. (Later Einstein himself would have occasion to balk at

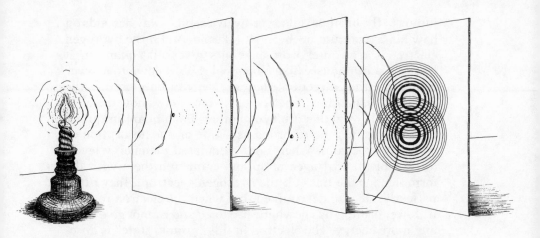

Thomas Young directed light onto a screen containing two tiny pinholes set closely together. Light passing through the pinholes fell onto a second screen. What Young saw was overlapping patterns like water waves which spread out and interfere with each other as they pass through gaps in a harbor wall. Light travels in waves, he concluded.

the implications of his own quantum discovery.) With Planck's and Einstein's discovery, the first of what have come to be known as the quantum paradoxes was born: Light and other forms of energy have a dual personality—at times behaving like a wave and at times like a particle! For the classical Newtonian physicist this made about as much logical sense as saying a drop of water is also a stone.

Bohr put together his two clues, the discrete quantum and the spectral lines, then applied them to Rutherford's atomic solar system and derived a surprising new model of the atom. What Bohr proposed was that the possible energy levels for an atom are "quantized," that is, they are fixed and discrete. An electron can move only in certain separated orbits, almost like a set of grooves drawn around the nucleus. An electron can't lose or gain energy in a continuous way, spiraling to higher or lower orbits; it takes in or throws off a packet (quantum) of energy and leaps from one groove to the other like a rabbit disappearing and reappearing discontinuously in various hats.

In the hydrogen atom, which Rutherford said had one electron in orbit, if a quantum packet of light energy is absorbed, the electron jumps to a higher orbit. In this case, whether it jumps to the orbit next door or twelve orbits away depends on how big a quantum of energy it takes in. And if the hydrogen electron is in an outer orbit and emits (gives off) a quantum, it leaps discontinuously (like a magical rabbit) down to a lower orbit. In either direction, the electron can only occupy the fixed orbits, nothing in between. No one knows how the electron travels from one orbit to another. Each quantum jump, up or down, will be recorded as a line on the absorption or emission spectrum. When Bohr calculated the energy levels (orbits) of the hydrogen atom and compared them with the formulae for the lines on the hydrogen spectrum, they fitted perfectly. He also determined that when the electron reaches its lower orbit, it has nowhere else to go and cannot give away any more energy. The electron in this "ground state" is absolutely stable. Stones or mountains made from atoms can therefore exist for eternity.

The electron (rabbit) jumps discontinuously from orbit to orbit by taking in or emitting quanta. The rabbit leaves a record of her magical jump as a line on the absorption or emission spectrum.

Bohr's theory was brilliant and had an immediate appeal for physicists of the time because it integrated the strange new insights of Planck and Einstein on the discontinuous quantum nature of energy into a traditional and familiar Newtonian framework of orbits around a central body.

HEISENBERG FEELS SURE

Following publication of Bohr's results, his new ideas were widely discussed in centers of learning throughout Europe. At the university at Munich, for example, the head of the theoretical physics faculty, Arnold Sommerfeld, gave a series of seminars on atomic structure which excited the imagination of two young students, Werner Heisenberg and Wolfgang Pauli, who spent their leisure hours arguing about these ideas.

It became clear to the two young men that despite the brilliance of Bohr's theory, there was something askew at its foundations. For one thing, experimentalists' improvements in spectral technology disclosed a finer structure of lines than the ones Bohr worked with. These new lines proved difficult to account for using the Bohr atomic model. But the young men went further than these experimental objections. They realized that Bohr had grafted the new quantum ideas onto older nineteenth-century notions like planetary orbits. While older physicists felt content with this hybrid, the compromise seemed unsatisfactory to the two students.

Sommerfeld noticed Heisenberg's interest, and even though the young man had not yet written his thesis, gave his student a theoretical problem to solve. He asked Heisenberg to identify the Bohr orbits corresponding to some recent experimental results. Sommerfeld then learned that Bohr would be giving a seminar in Göttingen and offered Heisenberg rail fare from Munich so that he could hear about the theory from its creator.

It is probable that Heisenberg was disappointed when he first heard Bohr lecture in 1922. From most accounts, Bohr was a particularly bad public speaker, given to spending an irritatingly long time trying to light his pipe and then mumbling in a soft voice so that only people in the first rows could catch the trend of his thoughts.

At the lecture, Heisenberg did something few students of the day would have had the nerve to do. He stood up and made objections to recent developments of Bohr's theory. Far from being offended, the older man was struck by this student's intensity and asked him to go walking that same afternoon. By the end of the day Bohr was sufficiently impressed to invite Heisenberg to visit him in Copenhagen.

Following that walk, however, Heisenberg was forced to return to earth; for the next three years he tended to his lectures and studies and wrote a thesis on hydrodynamics. But whenever he had time to spare, he would seek out his friend Pauli and argue about Bohr's theory.

As soon as his thesis was finished, Heisenberg began to concentrate his energies on the atom. What was really going on inside it? By the first months of 1925 his mind had become such a jumble of conjectures and mathematical equations that he actually fell ill. The physical cause was an attack of hay fever so severe that in May 1925 he took a leave of absence from the University of Göttingen, where he was teaching. He spent two weeks on the island of Helgoland, an area reputedly free from pollen.

The young man swam and took long walks, the perplexities of Bohr's atom never far from his thoughts. Physical activity and a change of scene appear to have cleared his mind, however, and he now saw the problem with fresh eyes.

Heisenberg realized even more clearly now that his confusion was caused by trying to hang on to traditional and trusted ideas of physics such as orbits and paths of particles. These worked very well in the everyday world of stones and cannonballs but were no longer appropriate at the frontier of the quantum. For a scientist trained in the great German tradition of mechanics, Heisenberg was questioning his catechism. If he threw away such notions, what could stand in their place?

Heisenberg recalled that his friend Pauli had once told him how Einstein (who had been greatly influenced by the logical positivist Mach) had founded his theory of relativity on what he called "observables." The famous physicist had taken as his starting point only those things that could be measured and observed; anything else was to be ignored as an unnecessary, metaphysical encumbrance.

Heisenberg decided that he would begin in exactly the same way—with the observable facts of the atomic spectra, the distances between the lines and the lines' thicknesses. (According to Bohr's theory, the thicker lines indicated energy levels which were "preferred" by electrons. The thickness showed that many, many electrons had jumped to those energy levels when large numbers of atoms were heated.) Everything else but these observables he would throw away, including the classical ideas of orbits and paths.

On his island, isolated from the university and its demands, Heisenberg made rapid progress. He took the numbers which describe atomic spectra (his observables) and arranged them in square patterns. Next he discovered a simple rule so he could manipulate these number patterns by treating a whole pattern like a single symbol in algebra. Heisenberg's progress at this point was a combination of trial and error guided by an overall intuition about the way things should go.

One night he was ready to make his first calculation on a simple case. The result is best told in his own words, from his autobiography, *Physics and Beyond*:

I reached the point where I was ready to determine the individual terms in the energy table, or, as we put it today, in the energy matrix. . . . When the first terms seemed to accord with the energy principle, I became rather excited, and I began to make countless arithmetical errors. As a result, it was almost three o'clock in the morning before the final result of my computations lay before me. . . . At first, I was deeply alarmed. I had the feeling that through the surface of atomic phenomena I was looking at a strangely beautiful interior, and felt almost giddy at the thought that I now had to probe this wealth of mathematical structures nature had so generously spread out before me. I was far too excited to sleep.[20]

The triumphant young scientist then walked to the southern tip of the island, climbed a large rock that jutted out into the sea, and watched the sun rise.

Heisenberg's results were quickly published and he soon found himself a star, lecturing to distinguished audiences across Germany. On one occasion, in Berlin, Albert Einstein sat listening in the audience.

After the lecture, Einstein invited the speaker to walk home with him. Heisenberg explained to Einstein how his quantum mechanics was founded only on observables, just as the theory of relativity had been in 1905. Einstein's reply was a surprise. He agreed that he might once have worked that way

Werner Heisenberg

but said, "It's nonsense all the same." Einstein—anticipating the approach Popper would take a few years later—explained that it was pointless to attempt to build theories on observables, for, after all, it was the theory itself which told physicists what could and could not be observed in nature.

Einstein's point was a deep and subtle one. Several months later, Heisenberg would have a dramatic occasion to make use of it himself.

Heisenberg rode the wave of success for a few months until a rival theory appeared on the scene and threatened to push his quantum mechanics into oblivion. The new theory of "wave mechanics" had its origins in the Ph.D. thesis of a French student and nobleman (later prince), Louis Victor de Broglie.

De Broglie knew that at times light behaved like a wave, bending outward from the pinholes in Thomas Young's experiments. But when it bombarded the surface of metal in the photoelectric effect studied by Einstein, it behaved like a particle. Light (and all energy) was both particle and wave.

De Broglie decided to take this idea one bold step further by asking: If waves of energy can possess a particle personality, then can particles of matter possess a *wave* personality? De Broglie suggested that the electron, for example, could at times behave like a wave and at other times like a particle. De Broglie's ideas aroused little attention when they were published, but his thesis, which was finished in 1924, was sent to Einstein, who realized the importance of de Broglie's "matter waves" and passed on his enthusiasm to a colleague in his old town of Zurich, Erwin Schrödinger.

WAVES OF UNCERTAINTY

Schrödinger was at the time in his late thirties, almost past what is usually considered the creative life for a theoretician. He had made no important contributions to theoretical physics; indeed, after his period of military service in World War I he had considered abandoning science and taking up philosophy. Nevertheless, he became excited at what Einstein told him. Suppose that the electron actually acted like a de Broglie wave when it was inside the atom; would it be possible to calculate the movement of this wave?

At the time, de Broglie's "matter" waves were purely speculation—with no experimental evidence to back them up. (A few years later, matter waves were actually photographed.) Schrödinger was able to show that if the single electron in a hydrogen atom was really a "standing wave," it would assume frequencies which were exactly equivalent to what Bohr described as discrete orbits with energies the same as those calculated by Heisenberg. By showing that the electron's energy levels could be wave patterns, Schrödinger solved the same problem of the new hydrogen spectrum that Heisenberg had solved with his collection of observables. The advantage of Schrödinger's system, however, was obvious. Schrödinger gave physicists a "picture" of the inside of the atom, whereas Heisenberg had given them only mathematics.

In 1926, the first of Schrödinger's papers appeared and was an immediate success. In April of that year, Einstein wrote him, "I am convinced that you have made a decisive advance with your formulation of the quantum condition, just as I am equally convinced that the Heisenberg-Born route is off the track."[26]

On April 2, Max Planck wrote, "I read your article the way an inquisitive child listens in suspense to the solution of a puzzle that he has been bothered about for a long time, and I am delighted with the beauties that are evident to the eye."[26]

To physicists it now seemed that Heisenberg's purely abstract algebra of the observables of atomic energy transitions had been a mere stopgap procedure on the way to a deeper theory. Even his former professor Max Born, who had initially contributed to the mathematical formulation of the theory, believed that Heisenberg's quantum mechanics would soon fade away.

But as physicists began to look more closely into the Schrödinger atom, the issue became clouded. To begin with, not only did the Heisenberg and Schrödinger theories both solve the same problem, it was also possible to show that the approaches were equivalent mathematically despite their conceptual differences.

Another complication appeared when the Schrödinger wave model was applied to atoms with more than one electron. In the hydrogen atom everything had seemed quite clear—electrons form matter waves and these waves set up standing patterns around the nucleus. The "wave function"—which is the name given to any solution to Schrödinger's equation—appears almost a direct representation of this matter wave, a mathematical picture of a concrete thing.

But what happens if an atom has two electrons? Does this mean there are two waves moving in space? At this point the safe, classical picture of the inside of the atom began to dissolve again. The wave equation for two electrons could not be written in three dimensions of space but had to be written in three dimensions for each electron—six dimensions. In the case of the three-electron atom lithium, nine dimensions of space were needed (3 + 3 + 3), and for beryllium, with four electrons, twelve space dimensions. Clearly Schrödinger's wave function did not directly correspond to anything physical at all, for it was expressed in abstract mathematical space.

The problem was solved by Born, who pointed out that Schrödinger's wave solutions were in fact waves of probability. This meant the Schrödinger equation didn't describe a physical wave so much as the probability of observing an electron (quantum particle) in a certain place. Electrons could be located by the simple procedure of firing a subatomic particle into a region of space. If the particle hit something that left a track on a photographic plate, then scientists knew the electron was there. Born said that where the Schrödinger probability wave picture was concentrated there was a good chance of finding an electron; where it was spread out there was still a chance but with a lower probability. Physicists began to realize the wave function equation was showing a far subtler aspect of reality than could be pictured.

Despite these problems, many physicists in 1926 preferred Schrödinger's equation because it conveyed at least some physical picture, while Heisenberg's matrix mechanics remained a coordination of quantum energy levels and transitions. H. A. Lorentz described the dilemma to Schrödinger in May 1926:

If I had to choose now between your wave mechanics and the matrix mechanics, I would give the preference to the former because of its greater intuitive clarity, so long as one only has to deal with the three coordinates x,y,z. If, however, there are more degrees of freedom, then I cannot interpret the waves and vibrations physically, and I must therefore decide in favor of matrix mechanics.[26]

S FUNCTION

P FUNCTION   D FUNCTION

The sketch shows a plotting of Schrödinger's wave function equations for the hydrogen atom. They represent three different levels of excitation of the single electron in the atom. Imagine the sketch is in three dimensions with the atom's nucleus located in the center of each figure. The dense, dark patch around the center shows the area where there is the highest probability of finding an electron. Where the dots are fewer, the probability is less. Quantum theory showed that theoretically an atom's electron might be found anywhere in space, even on the other side of a lead wall from its nucleus, though the probability decreases with distance.

Physics was torn between interpretations.

As physicists debated the rival theories, Heisenberg talked to Bohr about his misgivings over the wave function approach and Bohr invited Schrödinger to Copenhagen. To those who believe that scientists search after truth in a quiet and objective way, an account of the meeting between Bohr and Schrödinger in September 1926 may come as a shock.

Bohr greeted his guest at the Copenhagen railway station and immediately began to harangue him about the proper interpretation of quantum theory. The debate lasted long into the night only to resume early the next morning. According to Heisenberg, Bohr became fanatical: Schrödinger was too simplistic in trying to interpret the atom as waves.

Under the intensity of Bohr's attack, Schrödinger literally fell ill, but even then couldn't escape Bohr's relentless arguments. Schrödinger finished his visit exhausted but unyielding, and Bohr realized that a consistent interpretation of the quantum world was of vital importance.

Heisenberg took a flat in Copenhagen, and there in the winter and early spring of 1927 Bohr and he frequently talked all night. Heisenberg said later:

I remember discussions with Bohr which went through many hours till very late at night and ended almost in despair; and when at the end of the discussion I went alone for a walk in the neighborhood park I repeated to myself again and again the question: Can nature possibly be as absurd as it seemed to us in these atomic experiments.[20]

How could there be two such different theories—probability waves and a mechanics of observables—that could accurately predict experimental results? What was going on inside the atom that could explain how this was possible? Heisenberg's whole sense of an orderly and rational universe was at stake. But in the very despair lay the answer.

All the cherished ideas of physics and everyday common sense had been challenged by quantum behavior. Light is both wave and particle, electrons are both particles and waves. When a particle is shot from one point to another, it possesses no path in between. When the electron jumps from one quantum level of the atom to the next, it appears to have no existence in between. *Where* is the electron during a quantum jump? *How* can something be both a wave and a particle? Quantum theory seemed unable to answer such questions.

It was at this point that Heisenberg recalled what Einstein had told him on their walk home after the Berlin lecture. Was it possible that the answer to his questions already lay within the theory itself? But his quantum mechanical theory gave no answer. Was that somehow *the* answer?

Heisenberg was led by these musings to the "uncertainty principle"—a dramatic idea which gets to the root of the quantum world. He expressed this principle in terms of an imagined experiment using a microscope. The experiment has since been performed many times.

If we want to find out how the electron moves from point A to point B, why not use a very powerful microscope? Observe the particle at A, find out how fast it's going and in what direction, and calculate its path. Check this result a little farther along the path and continue until we get to B. This is exactly how a computer would track the approach of a missile and predict its point of impact.

Now the electron is very, very small. The wavelength of visible light is much larger than the electron, so with ordinary light it would be impossible to "see" the electron's position. For that we need light of incredibly small wavelengths—gamma rays.

Using a gamma-ray microscope it would be possible to pin down the position of the electron at point A. The next thing is to measure its speed and direction, or more accurately, its momentum (the momentum of a body is its velocity, including direction and speed, multiplied by its mass). But here Heisenberg saw that we run into a serious problem. With its small wavelength, the gamma ray is highly energetic, so that its photon (a photon is a quantum or particle of light) will have struck the electron a tremendous blow, speeding it up and knocking it out of its path. We know the position of the electron, but in the act of observing it we have totally changed its momentum (direction).

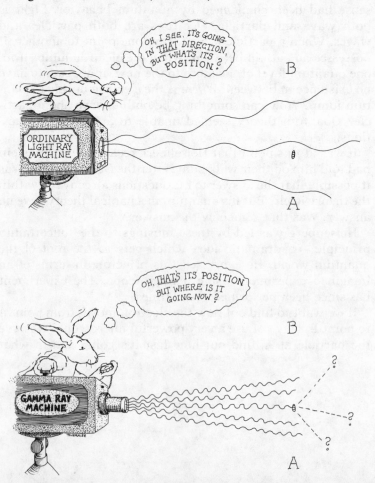

What if we reduce the power of the gamma rays so that individual photons do not change the momentum of the electron and throw it from its path? But if we use low-energy photons, this means we are using light of long wavelengths and we are back where we started: We can no longer locate the electron's position exactly.

Heisenberg's "microscope experiment" was a physical illustration of something he had discovered in the mathematics of his quantum mechanics itself: When the values of certain observables are measured, others become uncertain. The closer we try to measure the position of a quantum object, the more uncertain becomes its momentum. It seems the very act of observation or measurement changes the system.

Heisenberg's uncertainty principle showed that the actual properties of objects could no longer be separated from the act of measurement and *thus from the measurer himself.* Heisenberg had been given a vision of the looking-glass, and his work brought scientists to its edge.

Until Heisenberg, scientists had believed they could continually refine their experiments. Experiments always produce small disturbances that influence the result. The good experimenter will redesign the experiment, rebuild his apparatus, and progressively reduce unwanted external perturbation until he gets very close to the ideal situation where he has refined out all external influences and achieved purely objective results. In this ideal situation, the pre-Heisenbergian scientist is metaphorically seated behind a half-silvered mirror, a spectator to nature, observing things as they really are. With the uncertainty principle, as physicist John Wheeler was later to put it, the scientist smashed through that imaginary window separating him from nature.

In the fragments of that broken window lay the almost-completed map of the Newtonian paradigm. Heisenberg's principle signaled the final stages of the paradigm shift. Physicists were looking at the world with a whole new map. Spread out on this quantum map lay outlines of a paradoxical country.

The famous "double-slit" experiment illustrates the dilemmas scientists would now have to confront in trying further to map and navigate in this paradoxical paradigm.

Electrons are "elementary" entities; they can't be divided. Similarly, photons are single indivisible quanta of energy. From Thomas Young's experiment we know that if we fire a stream of photons (light energy) at a screen with two slits in it, the photons will interfere with each other and an overlapping wave pattern will form on the screen behind. We know from quantum theory that since matter is also waves, the same thing would happen if we fired a stream of electrons at the two slits. But what if we fired single photons or electrons one at a time at the screen?

According to classical physics, the individual electron or photon will travel through either one slit or the other. Each electron is a single, indivisible entity. There won't be any wave pattern because we shoot the electrons individually so that they won't interfere with each other. In actual experiments testing this idea, each time an electron goes through, scientists hear a single click on the detector screen, confirming that a single particle has landed. After a thousand or so such one-at-a-time firings, classical theory stipulates that behind each slit will be a simple scatter pattern, with about five hundred electrons recorded behind each hole.

The actual experiments show something else, however—something shocking. When scientists look at their detectors they don't find a scatter pattern, they find a wave pattern—just as if they fired a stream of electron-waves all at once!

There are two logical interpretations of this finding, and neither one makes any classical sense. In the first interpretation we say that each indivisible particle somehow manages to go through both slits at the same time and interfere with itself, contributing to the wave pattern on the screen, yet somehow registering as a single click. The second interpretation is even worse. We say that somehow each particle "knows" where the particles preceding it have gone and where the ones following will go so that at the end of the experiment they will all together have piled up in a nice wave pattern for the experimenter to puzzle over. Not only are these interpretations wild, neither one actually tells anything about what this entity (quantum) really is that scientists are experimenting with or how on earth it works. It is either very smart or it is indivisible and divisible at the same time.

Quantum scientists discovered that with their equations, particularly Schrödinger's probability wave function equation, they could very accurately predict what the patterns of large numbers of particles would look like but couldn't say much about what individual particles would do. This predictability of group behavior would eventually be one of the clues that would lead David Bohm to a looking-glass science.

For quantum paradigm scientists, probability thus became a "given." In other cases where scientists use probability, it is as a shortcut, a measure of ignorance. If you toss a coin, the probability is that it will come up heads 50 percent of the time. The more coin tossings are done, the more accurately this statement reflects the behavior of the group of tossed coins as a whole. However, such percentages are not much help in predicting which side the next coin will fall on. Scientists assume that to make that prediction you would need to know all the forces that go into any particular coin flip— thumb energy, air pressure, gravitational field. In a real situation we don't know all of these "hidden variables." There are too many of them to calibrate conveniently, so we rely on probability. Quantum probability is very different. Quantum scientists believe there are no hidden variables, not even extremely complex ones, in quantum reality. There are no unseen "reasons" to explain how individual quanta move from one orbit to another or appear to go through two slits at once. Scientists have only very accurate rules saying what large numbers of quanta will do, what their patterns will be. The rules of probability form a boundary on the quantum map.

COMPLIMENTS OF COMPLEMENTARITY

Bohr's concern with staying strictly inside that boundary made him react violently to Heinsenberg's microscope example for the uncertainty principle. Bohr approved of the uncertainty principle itself, believing it was an aspect of a deeper idea he called "complementarity." Complementarity meant the universe can never be described in a single, clear picture but must be apprehended through overlapping, complementary, and sometimes paradoxical views. Bohr found echoes of this idea in classical Chinese philosophy and the theories of modern psychology.

Bohr agreed with Heisenberg that it was important to show that the closer we measure the position of an electron, the more uncertain becomes its momentum. But he was less than complimentary about the example Heisenberg used to embody this idea. In fact, his reaction literally brought Heisenberg to tears. He complained that the microscope illustration assumed that the electron *had* an actual path, that at each instant it had definite momentum and position which was then disturbed by observation. It assumed that the electron really possesses definite properties and that it is only the quantum's interaction with the measuring apparatus that prevents us from measuring them exactly. Bohr argued vehemently that such classical holdovers must go. The electron has no path; it has no definite, independent properties. It is, in a phrase that would later become popular, not a "thing" but "tendencies to exist." At the smallest level of nature, Bohr said, we no longer find "things in themselves." He stood at the edge of a darkened mirror.

The picture of nature Bohr and Heisenberg eventually agreed upon and that became known as the "Copenhagen interpretation of quantum theory" is a far cry from the one that has stood since the time of the Greeks. The world we live and move in is composed of "things"—stones, houses, animals, ourselves. They appear to us as separate entities, with an existence of their own, their own properties of color, texture, smell, mass. Admittedly, there are other, less substantial entities like wind and water, but our basic picture of nature is of a world of things.

By the end of the nineteenth century it appeared that these "things" (including wind and water) were made of molecules and molecules were in turn made of atoms. Even Bohr's first model of the atom described a "thing" with its own unique properties, independent of any observer. But now Bohr and Heisenberg agreed that any property is, to some extent, a result of the act of measurement. As Bohr said, the photon depends on us to exist. And presumably the converse is also true. We also depend on it! There are no separate, independent objects. At the quantum level the world cannot be divided into independent parts, each exerting cause-and-effect relationships on the other, because, at the atomic level, everything is an indissoluble whole.

But what does "indissoluble whole" mean? The idea seems an enormous, mystical abstraction. Bohr's gaze into the looking-glass was a clouded one, and he pulled back from the edge, attempting to seal the border forever.

The new quantum mechanics, Bohr stressed, was doomed to the abstraction; it could make predictions but could no longer offer images and pictures. If a particular quantum mechanical experiment is performed, Schrödinger's or Heisenberg's theories will indicate how a dial will move or what lines will form on a photographic plate. These correlations will be statistically accurate—that's all the scientist can hope for. If we look at a group of particles arriving at point B (a detector) after particles are emitted from point A (a source), we can correctly calculate probabilities and patterns of what we find at B, but that's all; we cannot even say for sure that the particles detected at B were the same ones emitted at A, and we certainly can't say what path they took to get there. We can't even know what arrived at B, either. We have tracks on our equipment. But what they really are is uncertain.

Nowadays some physicists still argue that Heisenberg's uncertainty principle is only a statement about the limitations of observation (maybe the particle has both position and momentum but we just can't measure it). In other contexts, however, these same scientists are quick to embrace Bohr's idea of quantum wholeness, because it is the best argument available against the existence of hidden variables. Such a contradictory attitude illustrates the difficulty of finding one's way in the quantum paradigm map and might be called trying to have your quantum cake and eat it too. The problem is that it may be a looking-glass cake.

3. The First Expedition

We have seen how the second (quantum) expeditionaries reached the edge of the looking-glass through their probings of the universe in its smallest dimension—the atom. Though Albert Einstein had preceded these adventurers in this direction with his Nobel Prize–winning work on the quantum nature of light, he soon abandoned the trail and set off another way. He headed toward the universe in large. His visions and discoveries along this path between the stars led him to reject the theories of the quantum explorers and ultimately pull back, as they did—though for different reasons—when he came to the looking-glass threshold. He saw that threshold from one vantage point, they saw it from another. And, like people peering at a reality through different facets of a diamond, they could not agree on what they saw.

THE AMAZING ELASTIC CLOCKS AND RULERS

Einstein's theoretical excursion through the universe in its largest dimensions began with the special theory of relativity, published in 1905, when he was only twenty-six years old. It was the same year he did his work on light quanta. A decade later he expanded the special theory into the general theory of relativity. The two theories together are often referred to simply as "relativity."

Special relativity was born out of Einstein's concern over Newton's idea of absolute motion, that is, motion which takes place in absolute space and time. Absolute space can be visualized as a grid like that used by algebra students. During the Newtonian paradigm, the universe at large was pictured as an infinite three-dimensional grid and any movement could be measured and compared to any other movement in this imaginary grid the way you can compare two lines or curves plotted on the same graph.

The problem was that Newton had admitted that scientists might never be able to show by experiment that absolute motion really exists. It was rather as if we were fish in a sea of absolute space and time, unable to detect what we lived in. Ernst Mach had charged that Newton's idea of absolute motion was in fact undemonstrable and therefore metaphysical. For that reason it should be expelled from scientific theory. This argument seemed convincing to Einstein.

A second element in young Einstein's formulation of special relativity was born from his interest in light. Even as a boy, Einstein had been fascinated by light. At the time he was working on the problem of absolute motion, light—from gamma waves to radio waves—was pictured by scientists as electromagnetic vibrations moving in an elastic medium that filled empty space. This medium was called "ether." Light waves were said to travel through this invisible ether the way sound waves travel through air. The ether was an absolute "sea" in which everything moved. It was supposed that if an observer at rest in the ether measured the speed of a light ray going past him, he would get one value. But if the observer was moving as well, he would get another value. For example, if he was moving in the same general direction as the light ray, it would appear to him to be going slower. As a boy, Einstein is said to have wondered what would happen if he chased after a beam of light—faster and faster. When himself was moving at the speed of light, what would he see? The theory supplied an answer to this boyhood question. If he chased a beam of light with increasing speed, in the end he would run alongside it and see a stationary ripple in the ether. But this was clearly absurd. Such a thought experiment must have convinced Einstein at once that something was deeply wrong with the theory.

Experimental physicists, using the earth itself as a laboratory, tried to detect differences in the speed of light as the earth swept around the sun through the ether. They figured that if they measured the speed of light between two points, they should get different answers, depending on whether the light ray was moving in the same direction as the earth's movement around the sun or in some other direction. The results were surprising, however. No matter what direction they shot the light ray, when they measured the speed, it always came out three hundred million meters per second. This speed of light, called "c," appeared to be constant. For c to be constant the moving earth had to be always standing still in the ether. Ridiculous.

Just one year before the special theory of relativity, the Dutch physicist Hendrik Lorentz attempted to resolve this paradox. Lorentz came up with the ingenious argument that the earth really moved through the ether but a second effect exactly masked all attempts to detect its movement. Lorentz reasoned as follows: All experiments to measure the speed of light make use of clocks and measured distances. For example, speed is measured by timing how long it takes something to complete a given distance. But clocks and measuring rods are all made up of atoms, and the collection of atoms is held together by electrical forces. This much was clear by the end of the nineteenth century. Now suppose for a moment that the earth (or a laboratory) is at rest in the ether; all the atoms take up positions as the electrical forces between them balance. The physical objects—clocks and measuring rods—will have given atomic configurations, that is, specific rates of running and specific lengths. Suppose now that the laboratory begins to move through the ether. The electrical forces that hold the atoms of the clocks and rods together have to catch up with the moving atoms, so each atom is slightly shifted out of its position. The atoms become squashed together in the direction of their travel. The faster the laboratory moves, the more the atoms will be squashed. The net effect of these shifts is to make measuring rods contract and clocks run slower.

In other words, although the figure for c should be smaller, the effect of the contracted laboratory instruments is to overestimate c. Lorentz made a careful calculation of the change in length and time and discovered that it exactly canceled any change in the velocity of light.

Thus, Lorentz argued that although a laboratory may move through the ether, effects conspire to make this movement unobservable. The speed of light must always appear the same no matter how fast the laboratory moves. In a sense, Lorentz's argument was similar to the approach taken by Bohr in his early theory of the atom—a mixture of new and old ideas. Lorentz suggested the radically new idea that clocks and rods would change as a laboratory moved, yet he retained the old ideas of an ether and absolute motion.

Einstein felt the Lorentz argument was unacceptable. If motion in the ether could never be observed, then why not do away with the ideas of an ether and absolute motion altogether? Henceforth only relative motions would have meaning. Einstein took one additional step. Instead of considering the constancy of the velocity of light the coincidental accident that Lorentz suggested, Einstein now elevated it to a new principle of physics.

The theory of relativity contained certain surprises, for it asserted that observers who moved at different speeds would see events in quite different ways. What looked like a simultaneous event to one observer might appear like two quite distinct events to another. "Everything is relative," became the cry of the popularizers. "Einstein has done away with absolutes. Everyone sees a totally different universe." At first sight it may seem as if Einstein is saying something similar to Bohr, that each observer is creating a universe of his own. In fact, Einstein views the observer in a very different light from Bohr.

To Einstein the universe was not simply "relative." He firmly believed in a real, objective universe that existed inde-

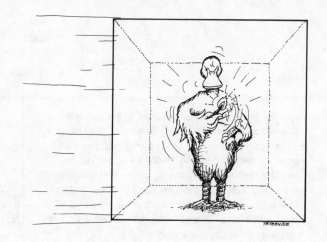

Einstein's insight included the relatively simple observation that observers moving at different speeds see events differently. The duck standing in the center of a room rocketing through space strikes a match. From her point of view, the light from the match reaches the front and back walls of the room at the same time. The rabbit on the asteroid, however, sees this event differently. For him, the match light reaches the back wall a short time before it reaches the front wall.

pendent of any observer. The act of observation in no way affected Einstein's universe.

But if each observer sees a different universe, how is it possible to describe that universe as "objective"? Einstein answered that the objectivity lay in the laws of nature. No matter how observers move, the same laws of nature always

From the point of view of the duck in the rocketing relativity room, her clocks run "normally" and her ruler measures the same as it always has. The rabbit, however, sees something strange happening. He sees the duck's clocks slowing down and the ruler contracting. Not only that, he sees this weird duck growing thinner. Instead of a universe made up of hard and definite "things," the rabbit realizes there is a deep elasticity to reality.

apply. Although each observer sees different phenomena and makes different measurements of the same event, the special theory of relativity contains mathematical rules for translating phenomena and measurements. It is rather as if each observer comes from a different country and speaks a different language, but by passing all their statements through a translator they can always be in agreement.

Suppose, for example, a group of scientists set up their laboratories in different rockets all speeding across the galaxy. From their rockets they carry out experiments and make measurements. Different laboratories may observe differing phenomena, yet the laws of nature they derive will be identical. For one thing, no matter where they are moving or how, they will all derive the same speed for light. The laws of nature are "invariant"; they do not depend on the state of motion of observers. This was the objective, real universe that lay behind the theory of relativity.

By following the path of the special theory of relativity, Einstein left the Newtonian paradigm behind. In a way quite different from that of the quantum theorists, his theory demonstrated that matter and energy cannot be uniquely separated. Whether something is matter or energy depends on the frame of reference. Put another way, Einstein found matter and energy are equivalent. He expressed this equivalence in the world-famous equation $E = mc^2$ that would forever be associated with his name.

Einstein showed how each observer would have his own space-time and how the rods and clocks of one space-time run at different rates from the rods and clocks of another. To

Lorentz this was an absolute effect; a clock ran slow because it moved with respect to the ether, and a clock at rest ran at "normal" speed. In special relativity the effect was more subtle: If observers A and B move at different speeds and carry clocks, then to A, clock B is running slow. But to B, clock A is running slow. The effect is quite symmetrical, just like the statement "A is moving past me" made by B and the statement "B is moving past me" made by A. Everything is relative in the sense that absolute motion does not exist. Only relative motion has meaning in Einstein's universe.

The fact that a moving clock runs slow is now well established. Certain elementary particles have only finite lifetimes and act like tiny clocks. The mu meson, for example, has a lifetime of around one millionth of a second; after this time, it will decay into an electron and two neutrinos. Mu mesons that are produced from high-energy cosmic rays traveling at speeds very close to the speed of light in a Newtonian world would be expected to move several miles in this millionth of a second. In fact the mu mesons travel much farther because their "internal clocks" run much slower and the mu meson's life is "stretched out" with respect to clocks on earth.

The discovery of the stretching out and contraction of phenomena moving relative to each other was almost a step into the looking-glass. In Einstein's universe, there are no hard "things" as such—no absolute particles, only relative ones. Everything is fluid and elastic. Einstein showed that the idea of an isolated rigid body was inappropriate to his theory. In the place of point particles and rigid bodies of Newtonian physics he introduced the "world tube" or history of a region of space. In essence the world tube is a basic object in the theory, for it cannot be analyzed in finer detail. A world tube can be pictured as a whirlpool in a river. From a distance, one can clearly see the turbulent water of the whirlpool and the slowly flowing river, but approaching closer we realize it's impossible to say where the whirlpool ends and the river begins—such analysis into separate and distinct parts fails. Bohr said something similar when he argued that a quantum experiment could not be divided up into absolute "parts." To their deep sorrow, the two scientists never were able to come together on this common ground of their theories—perhaps because this common ground lay in the looking-glass and neither would cross the boundary.

MORE THAN RELATIVE

Despite the success of his special theory, Einstein was aware of its incompleteness. If he was going to show that everything observers see is relative, but the laws of nature are invariant, he'd left one thing out of account—acceleration. When several observers glide along evenly in rocket ships going at different speeds, they are all equivalent in that they are relative to each other and absolute motion has no meaning. But what if one of them speeds up to overtake the other? Immediately, the observer in the accelerating rocket seems singled out. He is thrown back in his seat. He feels a force which none of the observers in the other ships feels. Doesn't the force of acceleration imply that, after all, there is some absolute motion? Special relativity, Einstein realized, was only half a theory; it stated that speed is totally relative, yet acceleration appeared to suggest there is an absolute background against which things move.*

Einstein puzzled over this defect in his theory for some time.

The beginning of a solution appeared one day in 1907 as he was working on a summary of the special theory of relativity. "At that point," he wrote later, "there came to me the happiest thought of my life."[64] He suddenly saw that a man in free fall from the roof of house would, paradoxically, not experience gravity. If he was carrying a briefcase and released it, the briefcase wouldn't fall but would remain where it was relative to him, though of course both man and briefcase would be falling relative to the ground.

This insight goes back to an experiment of Galileo. In popular mythology, Galileo is supposed to have dropped various weights from the Leaning Tower of Pisa and to have observed that they all hit the ground at the same time. While the Italian physicist probably never conducted this particular experiment from the Leaning Tower, he certainly knew that if we eliminate the effect of air resistance, no matter what a thing is made of, it will fall at the same rate of acceleration.

In the absence of air a styrofoam cup and a bowling ball dropped together will hit the ground together.

---

* In Mach's analysis of the failings of Newton's absolute motion, he makes what amounts to a looking-glass proposal. This will be clear when we come to David Bohm. Mach thought that inertia—which is the resistance to acceleration—is caused by *a body's interaction with the rest of the universe*. Thus, when a body is accelerated by an external force, it is also reacting with all other matter in the universe. *Everything is tied in with everything else.* Though Einstein greatly admired Mach's analysis of Newton, his own solution to the acceleration problem lay down another path.

This example can be refined even further. Suppose a laboratory is placed in a rocket standing still at the launching pad on earth. Inside the scientist drops a number of different weights. They all hit the ground at the same instant. He swings a pendulum and it ticks away the seconds. All his measurements indicate a gravitational pull inside the rocket equal to that of the earth.

Suppose now the rocket is accelerated in outer space far from the earth at 32 feet per second per second (the same rate at which objects fall in earth's gravity). The scientist drops various objects and they all fall to the floor at the same rate (but in another sense they remain in space weightlessness where they have been released and the rocket floor accelerates up to them). The scientist swings a pendulum and it ticks away the seconds. All his measurements indicate a gravitational pull equal to that of the earth. Or do they indicate an acceleration of 32 feet per second per second? In fact the scientist has no way of telling if he is at rest on earth or being accelerated through space—that is, unless he looks out of the window and compares his motion relative to the motion of something else like a planet or another rocket ship.

Examples like these convinced Einstein that gravity and acceleration must be indissolubly tied together. He was convinced that Galileo's results with falling weights were no simple accident of nature but involve a profound insight into a new law. Just as Einstein had turned speculations on the velocity of light into the cornerstone of his special theory of relativity, he now made the interconnection between gravity and acceleration the foundation of his general theory of relativity.

Gravity and acceleration are mathematically equivalent, Einstein realized. But acceleration has to do with rates of change of velocity, that is, with the path of a particle. It is therefore a geometrical concept. Gravity must therefore also be a geometrical concept!

Newton had said the moon accelerates in its orbit around the earth because, in effect, it's always falling, pulled by the earth's gravity. But if gravity and acceleration can be considered the same thing, Einstein reasoned, what looks like the force of gravity pulling things in curves (or orbits) around planets is actually acceleration. Therefore, if things accelerate in curves, he realized, it must be because space itself is curved. He called these curves in space-time "geodesics." A

The bell jar stands on the table and Alice lets go of her hourglass. It crashes to the floor, and she knows this is because of gravity. Next, she is traveling in her jar through empty space. She wears magnetic shoes, so she is anchored to the floor. This time when she lets go of her hourglass, the same thing happens. But now she doesn't know for sure what caused it. Did the hourglass fall because it was acted on by a gravitational field or did the floor rise to meet the hourglass because her jarship was accelerating? (Alice knows one or the other must have happened, because if she were still moving through space in uniform motion the hourglass would simply have floated when she let it go.)

rocket ship which is not being pushed along by the force of its propellants will move along the natural curves or geodesics in space-time. The more curved space is in any region, the more the rocket will accelerate.

We could also say there is a greater pull of gravity. But that means putting all this the other way around—that because bodies such as planets exert a gravitational force they warp space-time into curves. Imagine a stretched, thin sheet of rubber onto which are placed several different-sized balls. The balls will sink into the sheet and warp it. This is what the gravity of celestial bodies does to space. Now if you were riding a marble (or rocket ship) and wanted to get from one wooden ball (celestial body) to another, the shortest distance or straight line would be a curved line following the warping of space. Your marble rocket ship would accelerate as it followed the steepest curves near the heavier balls. To express the curved-space idea mathematically, Einstein used the fanciful geometry of Georg Friedrich Reimann. We should note that Einsteinian space is, of course, not in two dimensions like the rubber sheet. Actually, it's in our familiar three dimensions plus time (because of the necessity to include time as a dimension in which observers move).

But Einstein did not believe the geometry of space-time was *caused* by the gravity of heavenly bodies. As far as he was concerned, gravity is itself the geometry or curvature of space-time.

For Einstein, celestial bodies like the sun or the earth are themselves intense curvatures of space-time; they in turn warp (or are warped by) the space around them.

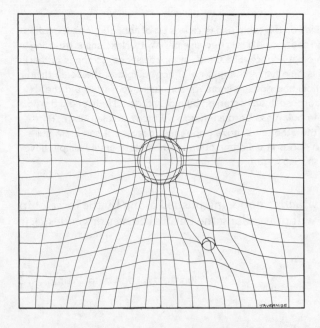

General relativity provided a new theory of gravity, based on curved space-time, to replace Newton's gravitational force. The differences in the predictive value of the two theories were small but sufficient to account for the anomalies that had been observed in the orbit of the planet Mercury. That anomaly, along with the paradox of the infinite light energy inside a "black box," was the last classical physicists believed they needed to solve before the Newtonian paradigm would be complete. Einstein's revolutionary solution to both these problems brought an end to that Newtonian dream.

Dramatic confirmation of Einstein's relativity theory came when the bending of starlight by the gravitational field of the sun was first observed during the solar eclipse of 1919. Since then, accurate atomic clocks have been monitored and found to run at different rates in different gravitational fields, for example at sea level and on a high mountain. Thanks to artificial satellites, laser distance determinations, and accurate atomic clocks, the various predictions of general relativity are now well established. Even more exotic predictions of the theory, such as "black holes" caused by the intense warping of space, have been provisionally accepted by the physics community.

Ultimately, Einstein hoped to take relativity further by eliminating the ideas of other forces such as electricity and magnetism along with matter and replacing them all with a geometry of space-time. He envisioned a "unified field theory" in which matter would appear as concentrations or humps in a universal field. Though he was never able to achieve this theory, his path in pursuit of it took him close to the looking-glass. Einstein believed the universe is whole, that everything is ultimately equivalent to and transforms into everything else (depending on the relative motion of the observer): Matter is equivalent to energy, gravity to acceleration, space to time. It is all one unified field.

NO DICE: THE EXPEDITION LEADERS QUARREL

Both Einstein and Bohr agreed that there was a wholeness to the universe, but the theories that led to this insight contradicted one another. Relativity is based on the idea of continuous fields, while quantum theory reveals nature as discontinuous (those quantum jumps). In quantum theory the idea of a path or trajectory is abandoned, while in relativity the geodesic has a special position. Einstein provided a set of mathematical rules, or transformations, so one observer could translate what he observed to any other observer. Individual observers may see phenomena in slightly different ways but, by using the translation rules, they will always discover the same laws of nature. This implies that the universe is objective and deterministic. It's independent of observers, and the observers discover laws of certainty not probability. Quantum theory, on the other hand, had unveiled an indeterministic, probabilistic universe and asserted that observation could not be independent from observers. Einstein's theory provided a clear cause-and-effect picture of the relationship between observer, observations, and the elements of the universe at large. Bohr's Copenhagen interpretation of quantum theory said that in the universe in small there could be no definitive picture of cause and effect. There could only be probability waves and correlations of observations.

Einstein insisted that quantum theory's lack of clear cause-and-effect pictures meant it must be "incomplete." He attempted to demonstrate this incompleteness in a number of thought experiments. A well-known one involves the disintegration of an atomic nucleus.

Certain nuclei are inherently unstable and can achieve relative stability only through internal rearrangement of their elementary particles. This takes place by the emission of a particle from the nulceus. One form of this radioactivity is beta decay, the emission of an electron.

The process can be described by the rules of quantum theory in the following way: Schrödinger's equation for the electron inside the nucleus is written down and then solved. The solution gives the wave function at successive time intervals, beginning with the electron totally contained inside the nucleus. At this initial time, the wave function is strongly peaked or localized inside the nucleus, but a little while later the tail of the wave function will have spread or "leaked" out. The result is something like what would happen if you tipped a carton of ice cream onto a kitchen table. At first the ice cream would lie in a solid lump, but as it heated up a thin layer of milk would spread and the ice cream would slump.

As time passes, the wave function peak diminishes in size and the "tail" of the wave function thickens and spreads out. Eventually, when a sufficiently long time has passed, the wave function has spread out over all space and there is an equal probability of finding the electron in any location. Actual experiments have verified that even when barriers are placed in the way of an electron it will in fact leak out and be found outside the barriers at just the probability indicated by the Schrödinger equation. The Schrödinger equation therefore tells us that to begin with, there is 100 percent probability of finding the electron inside the nucleus, a little later there is a small probability of finding it outside, after more time has elapsed there is an even chance that the electron is outside the nucleus, and after sufficiently long time a scientist would be almost certain to find the electron outside.

But this picture of electron emission, this mental image of something which leaks from the nucleus in a continuous way, is not at all like what occurs in reality. Experimentally, a physicist surrounds the atom with detectors and waits for one of them to register a disintegration. In a typical experiment, a detector may register the emission of an electron after half an hour. The experiment will be repeated and the distintegration may not take place for two hours; on the next occasion it may occur after ten minutes. If the disintegration of a particular isotope is measured over a large enough number of events, it becomes possible to draw a picture of the probability of an electron's being registered in one of the counters after a given time has elapsed. This probability distribution is exactly the same as that calculated from Schrödinger's equation.

For Bohr this is all quantum theory can and should be doing; it cannot predict the exact instant when the atom emits the electron; all it can give is a way of calculating the probability of such an event, and if this probability agrees with experiment, then quantum theory has been shown to be correct. But such an explanation did not satisfy Einstein, who felt the theory should go further and predict exactly when the isotope will disintegrate. He pointed out that just before the counter registers, we know with 100 percent certainty that the electron is inside the atom, but as soon as the counter clicks we know with 100 percent certainty that the electron has left the nucleus. In other words, the real phenomenon is a discontinuous one while the picture given by Schrödinger's wave function is of a smooth, continuous process. A micro-

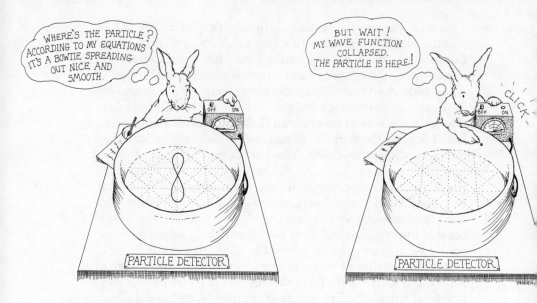

second before the counter clicks, the quantum mechanical wave function is spread out all over space; a microsecond later we know the electron has registered and the wave function has "collapsed." Yet nothing in the Schrödinger equation will explain such a collapse.

To Einstein and others, this "collapse of the wave function" was a clear indication that something was missing from the Schrödinger equation. Possibly if these missing terms or "hidden variables" were added, then the collapse of the wave function would be accounted for and the disintegration of an atom explained in exact detail. Bohr and his successors countered by claiming that hidden variables were an illusion and that quantum theory is a complete description of nature.

For Einstein, however, the universe just had to be objective and deterministic. In his autobiography he wrote that as a boy of twelve he had longed to free himself from a universe of the "merely personal" and had come to believe that "out yonder there was this huge world, which exists independently of us human beings."[41] He insisted that God "does not play dice." Where physicists resorted to statistics and probabilities it must be out of ignorance, as in games of chance. Underlying processes must be totally causal and independent of observers. He was convinced that there must be as yet undiscovered phenomena at work beneath the surface of quantum events, a hidden physics that when discovered would account for why the electron appeared at one time and not another.

Einstein was doubtless supported in this belief by a success he had in 1905 when he discovered hidden variables that lay behind "Brownian motion."

Tiny particles of dust can be seen dancing in a beam of sunlight. If a dusty room were to be totally controlled so that no breath of air stirred, these sunbeam particles would still perform their dance. Early in the nineteenth century the Scottish doctor and botanist Robert Brown noticed a similar motion in pollen grains suspended in water. He assumed it arose out of some mobility of the pollens themselves and was amazed at the "very unexpected fact of seeing vitality being retained by these 'molecules' so long after the death of the plant."[8] In fact, it had nothing to do with living pollen but could be seen with anything tiny—small particles of rock, dust in the air—an endless, random movement.

The reason for this motion was finally solved by Einstein, using a hidden-variable theory. Einstein proposed that the random motion of the tiny particles was a result of their bombardment by invisible molecules—water molecules or molecules of air. Each invisible molecule moves very rapidly and gives a tiny kick to the grain of pollen or speck of dust. The accumulation of many, many kicks produces a random jiggling of the grains.

At one level, the Brownian motion appeared to be totally random and probabilistic, yet as soon as one moved to a deeper level, the molecular level, the motion turned out to be totally deterministic. Could the same thing apply to atoms? Could the random processes of quantum theory result from a large number of deterministic subquantum events?

In 1935, Einstein and two young colleagues, Boris Podolsky and Nathan Rosen, had another crack at trying to show that quantum theory is incomplete and requires an additional theory of subquantum hidden variables. The three scientists proposed a thought experiment (called EPR for short) which was later refined in its details by David Bohm.

In the EPR, an atomic particle P disintegrates in two and the two halves, A and B, fly off in opposite directions at high speed. According to the laws of both classical and quantum physics, there is a correlation between the momentum of A and B and also a correlation between the position of A and B. This means that whatever we find out about A gives us corresponding information about B. Now Heisenberg's uncertainty principle stipulates that as we measure the position of particle A, its momentum becomes uncertain. But, Einstein and his colleagues asked, how could a measurement of A possibly affect particle B, which is flying towards the other side of the laboratory? Since the momentum and positions of A and B are correlated, then by acting on A, it should be possible to deduce something about B without actually measuring it. Suppose then that we measure the momentum of A. This also gives us the momentum of B. Without violating the uncertainty principle, we can then measure the position of B. But now we have a paradox. We haven't violated the uncertainty principle, yet we have managed to obtain both the momentum and position of B, a feat which the uncertainty principle says is impossible!

Bohr's reply to EPR was subtle, but, in brief, he insisted that Einstein's argument was invalid because it violated an important principle of the Copenhagen interpretation, the principle he had insisted on with Heisenberg. Quantum mechanical systems are *undivided wholes*, a single system. Einstein had assumed that particle B was "separate" while a measurement was being performed on A. According to Bohr, a measurement on any part of the system would change the whole system and the notion of disturbing one particle to obtain information about its distant partner was a fallacy. No matter how many times the scientist runs back and forth between the two particles A and B measuring their position

Einstein assumed a measurement on a particle here could not affect a particle there. Bohr said it could, and that it was wrong to talk about things being here or there as if they existed without an observer present to take a measurement.

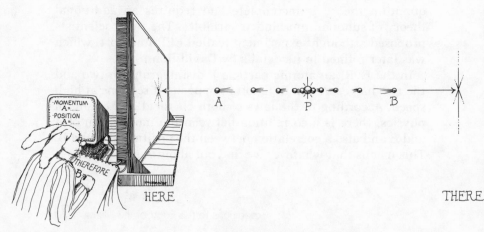

HERE   THERE

EPR EXPERIMENT

Niels Bohr, left, and Albert Einstein during one of their meetings.

and momentum, total certainty is never possible. For in each measurement the observer selects a new quality to measure.

At this point, as far as Bohr was concerned, quantum theory was vindicated and hidden variables had had their day. He was supported by a mathematical proof from John von Neumann which purported to show that hidden variables constructed by analogy to Brownian motion would be inconsistent with the mathematical tenets of quantum theory. Bohr said people looking for an alternative to quantum theory were like those who say, "We may hope that it will later turn out that sometimes 2 + 2 = 5."

In the late 1920s and early 1930s, Einstein and Bohr met on a number of occasions. Einstein said in a letter after one of these meetings that he never felt closer to anyone than he did to Bohr, and Bohr indicated a similar experience. Nevertheless, their long discussions on quantum theory, Einstein's relentless thought experiments, and Bohr's increasingly impatient replies eventually drove the men apart. Kuhn says that when a scientist refuses to accept the new paradigm after a shift has occurred, his colleagues begin to think of him as no longer practicing science. Einstein suffered that fate. He was revered for his early work but pitied for his inability to accept the new quantum science. He was one of those hoping that 2 + 2 would equal 5. For his part, Einstein wrote bitterly in 1928: "The Heisenberg-Bohr tranquilizing philosophy—or religion?—is so delicately contrived that, for the time being, it provides a gentle pillow for the true believer from which he cannot very easily be aroused. So let him lie there."[26]

Perhaps the greatest tragedy of that broken friendship was how close the men were on an important point in their theories. Both had discovered wholeness, but for each it meant different things, and not very clear things, unfortunately. By insisting that the laws of nature are objective, Einstein barred himself from crossing the threshold into the looking-glass of wholeness. Bohr set up his own barrier by insisting that no picture of this wholeness is possible, that the paradoxes and probabilities of the quantum theory are all there is to know.

4. Crisis at the Brink of the Looking-Glass

Since the days of the unhappy arguments between Bohr and Einstein, the quantum and relativity theories have both become firmly entrenched in modern physics. They are its current paradigm. Physicists accept the theories as unquestionably true and believe that rapid advances are being made in unifying them. Many physicists are further convinced that when this is done they will have also resolved most of the complex theoretical puzzles of their science. For most modern physicists there is no paradigm crisis.

GRAND UNIFICATION

Backed by all the research funds and public support an established paradigm can claim, contemporary scientists have been devoting most of their efforts to formulating and proving what is called "the grand unification theory"—a conceptual attempt to unify all the forces of nature.

Physicists believe there are four basic forces which act on

matter. The first two have been known to science for some time: electromagnetism, which most people are familiar with as the force that turns electric motors and carries TV signals; and gravity, which causes things to drop. The other two forces are more recent discoveries: the strong nuclear and weak nuclear force. The strong nuclear force binds protons and neutrons together in the nucleus. The weak nuclear force is responsible for the decay of certain elementary particles and seems to operate only when particles are extremely near each other.

The object of the grand unification theory is to provide a way of mathematically translating these four forces into each other so it can be shown how they all evolved from, and are expressions of, a single force. This would mean physicists had discovered a single unified field of force which accounted for everything—the very field Einstein hunted for in his later years. The grand unification theory assumes, however, that everything, including the four forces, is fundamentally quantum in nature and obeys the statistical laws of quantum mechanics. Thus, were Einstein alive he would probably find himself rejecting grand unification even if it were to succeed in bringing relativity and quantum theory together—because for Einstein any theory based on statistics must be incomplete.

In their quest to accomplish grand unification, physicists have made extensive use of experimental results coming from particle accelerators—huge electromagnets which drive particles at terrific speeds and smash them into each other in order to discover what, if anything, is inside. Recently this is reported to have yielded success in discovering the quantum which carries (or, more accurately, *is*) the weak nuclear force. Scientists now believe they can show that the weak nuclear force is the electromagnetic force in disguise. If true, this is a major step in unifying the four forces.

Over the years, particle accelerators and other probings by physicists have turned up a host of new entities in the subatomic realm. In the 1920s when Bohr and Heisenberg were struggling over interpretation of the new quantum theory, it looked as if matter were composed of only two elementary particles—the electron and proton. By the early 1960s, however, scientists had discovered a virtual "particle zoo" that included nearly a hundred elementary entities. And discoveries are still coming.

Experimental physicists have not been altogether unhappy with this situation, since it has kept them well supplied with new puzzles to solve. Many of the new particles have been discovered by applying implications of quantum theory. As the new particles turned up, they created new possibilities for still further things to discover and also new uncertainties about how all these things fitted together. The process has been reminiscent of Kuhn's description of how puzzle solvers of "normal science" extend and clarify the map of a paradigm, unwittingly generating new ignorances at the same rate they develop new understanding.

This, however, does not concern most practicing physicists. As far as they are concerned, quantum theory works. Again and again, its predictions are confirmed. In the normal process of research, quantum theorists don't really need to understand why quantum theory works but only how to apply its rules. This is particularly true for experimentalists who use theory as their guide. It's also true for a certain band of theoreticians who are not revolutionaries and don't mean to be but perceive their task as one of extending the prevailing theory by applying it to new problems.

This is understandable. Suppose a cook wishes to make a birthday cake with the aid of a cookbook. All that need be done is follow the directions in the recipe. The cook needn't concern himself with the biochemistry of starches, the mechanism of the oven, the coefficient of expansion of air trapped inside each tiny bubble of cake mixture, or the effect of heat on flour. To produce a good cake the cook only has to know the "how" of the recipe, not "why." Provided the cake turns out to be successful, there is no need to rewrite the cookbook.

In the same way, the quantum scientist needn't bother about long discussions about the interpretation of quantum theory. He simply has to apply the rules properly and calculate the answer. Since the answer has always proved to be correct, it seems as foolish to change quantum rules as it would to change a good cookbook. Indeed, if the value of a theory is judged by the numer of productive puzzles it presents, quantum theory must be accounted one of the most successful theories in science. Puzzle after puzzle has yielded to it. To take just one example, by 1928 Heisenberg's former professor Arnold Sommerfeld had applied quantum theory to the electrons which moved inside a metal. Sommerfeld took up an earlier hypothesis that these outer electrons were not in fact bound to individual atoms but moved freely. They were an "electron sea." By applying Schrödinger's equation to this electron sea, Sommerfeld was able to calculate the correct energy distribution of electrons in a metal, and, in the years that followed, the theory was extended to resistance, super-conductivity, metal catalysis, and semiconductors. By the 1980s, solid-state physics had become the largest branch of physics, all of its impressive work based on quantum theory.

Yet, because of the paradoxical nature of the principles of the quantum, the cake which quantum theory bakes has sometimes proved strange.

Typical of this strangeness is the discovery that when some elementary particles are smashed in an accelerator, they divide into a number of other particles, continuing to divide until *they divide back into themselves.* Something like this also happens to quarks.

A looking-glass particle cake. The diagram shows an example of a particle (proton) dividing up into a number of particles (neutral, positive, and negative pions; neutrons; an antineutron and an antiproton)— ending up, after all these splits, as itself.

The "quark" (which appears in a ditty from James Joyce's *Finnegans Wake*) was originally proposed by physicists to solve the problem created by the seemingly inexhaustible proliferation of particles. Scientists since the time of Democritus have believed that beneath nature's complexities must lie an innate simplicity. At the bottom of the universe there must be only a very few elementary types of entities and from these everything else is constructed. But modern physicists had nearly a hundred elementary entities. That was hardly simple. Einstein had said, "God is subtle but He is not malicious." However, particle physicists began to suspect there might, in fact, be a touch of maliciousness in creation. But quark theory came to the rescue by proposing that all those different elementary particles were themselves made up of still more fundamental entities—quarks. As quark theory matured, no quarks were actually observed, and physicists determined there was a very peculiar reason for this. First, they found that there was an unusual force holding the quarks together. They called it the "glue" force or "superstrong nuclear force." When quarks lay close together in a particle such as a meson, the glue force looked fairly weak. But when scientists applied energy to try to pry the quark out, the glue force got stronger and stronger. Physicists realized that when so much energy is applied that quarks *have* to break apart, the energy being applied to the quarks is instantly released and creates new quarks. These then combine with their freed siblings. Thus in the same instant a quark is divided from a meson, it combines with a newly created quark to form another meson. The net effect is that when a meson is split, new mesons are formed and no free quarks are seen.

The quantum mechanical cake seems a looking-glass cake—a wholeness which cannot be divided.

Looking-glass scientist David Bohm criticizes the current effort of quantum theorists to achieve a grand unification theory because he believes the pursuit of the ultimate particle, the ultimate quantum, or the ultimate force makes a fundamental error—it assumes that the universe is made up of parts. By pursuing nature as if she were made of parts, scientists have found parts. Bohm thinks, however, that no ultimate or "grandfather" part will be discovered, only more and more parts—parts which will keep elusively dissolving into themselves.

Another looking-glass scientist, Rupert Sheldrake, has

pointed out that even if quantum physicists are successful in deriving the ultimate force, the one which supposedly was responsible for the big bang that gave birth to the universe 15–20 billion years ago, they will still be faced with an insurmountable problem. The ultimate seed force would necessarily be the first law of nature. But where did this first law exist *before* the big bang? The answer to that question, Sheldrake says, must be metaphysical. He believes that hard-nosed mechanists who claim to reject all metaphysical assumptions are unwilling to recognize that their own theory is based on the vague philosophical idea that an eternal law of nature can exist even before nature herself is said to exist.

Some physicists make another complaint against the grand unification approach. They worry that trying to understand nature by smashing particles together is like trying to understand time by smashing clocks together.

One might picture the aim of grand unification as a theoretical engineering feat, an attempt to hold the whole "zoo" of mechanisms of particles and forces together in one structure. Though modern physics declared the end of the great Newtonian machine, grand unification physics seems to be trying to erect a quantum machine in its place. In the excitement of discovering new particles and solving new puzzles with quantum theory an important element seems lost. In the accumulating piles of knowledge, it is easy to forget that the theory stipulates we can never really know what is going on inside the atom. Quarks, gluons, and other subatomic quanta about which we seem to know so much are "really" only correlations and mathematical abstractions which sometimes leave tracks on experimental equipment. Our knowledge has been made by shadows.

So while there is the undoubted success of quantum theory in proposing and solving puzzles, there remains a disturbing and fundamental ambiguity at the heart of the theory. The ambiguity can be highlighted as three "problems" which suggest that although most physicists wouldn't admit it, there is in fact a "crisis" in the current paradigm.

PROBLEM 1: THE QUANTUM-RELATIVITY DILEMMA

So far the grand unificationists have not made much progress in giving a quantum description to the force of gravity and unifying it with the other forces. As a result, quantum theory and relativity remain as far apart as they were in the days of Einstein and Bohr. The eternal laws of the two theories seem irrevocably at odds.

Yet each theory in some sense relies on the other. Relativity, for example, is about the structure of space-time and makes use of clocks and measuring rods. The most accurate clocks are atomic clocks and the most accurate measuring rods are beams of light—both of which are quantum mechanical in nature. Conversely, quantum theory, as Bohr showed, relies on the observer who makes measurements on atomic systems using large-scale apparatus. But this very apparatus is ruled by the laws of relativity.

If the phenomena covered by relativity and quantum theory were always widely separated, then the desire to unify the theories would be only an aesthetic one, motivated more by a need for tidiness in science than immediate necessity. But in the last decade it has become apparent that there are phenomena in which the two theories overlap. The effect is like having two sergeants trying to drill the same group of soldiers at the same time.

One of the most dramatic examples of this overlap was discovered by the brilliant mathematical physicist Stephen Hawking, and arises from two assumptions, one from quantum theory, the other from relativity. Let's take the quantum mechanical assumption first.

Heisenberg's uncertainty principle tells us that not all the properties of a system can be known exactly. If a certain property is measured precisely, then another will become uncertain. An example of this indeterminacy is in the measurement of energy and time. If the energy of a system is measured over a reasonably long time period then it can be known very accurately, but if we attempt to measure this energy in shorter and shorter time intervals, it becomes uncertain.

Quantum physicists believe that this uncertainty leads to the following curious state of affairs—that the energy of any system is not strictly constant but from instant to instant fluctuates about an average value. If the system is observed in smaller and smaller time intervals, its energy appears to be in a greater and greater state of flux, increasing and decreasing at random but averaging out to a constant value. Since energy is always quantized (created in bursts), this must mean that the system is constantly creating and absorbing energy quanta. Now, energy and mass convert into each other according to the equation $E = mc^2$, so the system will also be creating and annihilating pairs of particles and antiparticles. Since these particle-antiparticle pairs are created for only a tiny fraction of a second before being reabsorbed, they are never directly observed in an experiment and are called "virtual particles." They are like money borrowed from the bank and repaid on the same day so the balance on the bank statement is unchanged.

The second assumption Hawking dealt with was a relativistic one. Einstein in the general theory taught that matter causes space-time to curve and this curvature is felt by nearby bodies as the force of gravity. According to Einstein's equations there is no reason why a sufficiently concentrated piece of matter should not cause space-time to curve so dramatically that it actually comes back on itself. Such a phenomenon is known as a black hole. Its curvature is so intense that nothing can escape from its center and everything that enters the black hole is torn apart by the forces of gravity. For each black hole there is a well-defined "point of no return" or "event horizon." Just to one side of the event horizon, particles can still escape, but a fraction to the other side, particles are doomed to extinction.

These facts, about the event horizon and black hole and the way particle-antiparticle pairs are constantly created and destroyed in quantum energy fluctuations, are well known to scientists, but no one had thought of putting them together before. Hawking asked what would happen if a virtual pair of a particle and an antiparticle were created at the event horizon. According to quantum theory, the pair would move apart for an instant, then recombine to form a quantum of energy. But what if one of the virtual particles happened to wander across the event horizon during this process? According to black-hole physics, this particle could never return to the outside world and its partner, unable to locate its mate, could not annihilate itself. In other words, if one of the virtual pair were to cross the event horizon, the other would be forced to appear as a real particle. The effect would be like borrowing money from the bank and having the bank suddenly close before the money could be repaid.

The effect which Hawking predicted was that, based on the principles of quantum theory and relativity, elementary particles will be continuously created around the event horizon of a black hole. (So far this effect has not been actually observed; indeed, no one yet knows for certain if black holes really exist.) Hawking's conclusion was that quantum processes and the curvature of space-time postulated by general relativity are therefore intimately interconnected and it is more important than ever for these two theories to be properly meshed.

PROBLEM 2: WHAT MEASUREMENT AND WHO SAYS?

What is the relationship of the observer to the observed? Put another way, what is happening when the observer takes a measurement? This question is crucial to quantum theory, but the theory itself has precious little to say about it. Some scientists feel it may be well and good to develop a grand unification theory based on particle experiments, but what does such a theory mean when the precise effects that the experimenter is having on his results are uncertain?

Ironically, Erwin Schrödinger, a founding father of the quantum paradigm, felt keenly that this was a particular failing of the theory and crystallized his reflections in the form of a thought experiment that became known as "Schrödinger's cat." It is a dramatic restatement of the atomic-decay experiment which so troubled Einstein. Schrödinger's version of the collapse of the wave function problem is formulated so as to focus on the problem of the role of the observer.

In the Schrödinger cat experiment a live cat is put in a box along with a sealed capsule of cyanide. (Let's make clear that this is only a hypothetical experiment and no physicist has ever suggested it should be done on a real cat.) Along with the cyanide capsule is a random device with a 50:50 chance of being activated. If the device is set off, it will strike and break the cyanide capsule, release the gas, and kill the cat; if the device is not set off, the capsule will remain intact and so will the cat.

Schrödinger's thought experiment employed a quantum mechanical device to smash the capsule, but, to clarify matters, let's begin with a familiar "classical" device—a roulette wheel. The wheel can be wired so that if the ball falls into one of the red slots it will trigger the breaking of the capsule; if it

Erwin Schrödinger

falls into a black slot nothing will happen. We set the wheel
rolling, pet the cat, and lock the box.

We return an hour later and ask ourselves what has occurred. Clearly, until we unlock the box we do not know if the
cat is alive or dead, although there is a 50:50 chance that
either event will have taken place. One thing we do know,
however, is that the cat is either alive or dead and that no
other outcome is possible. Opening the lid of the box will give
us additional information, but in no way will it affect the
outcome of events; the cat will be just as dead or alive
whether or not we open the lid of the box.

Suppose now we replace the roulette wheel with the disin-
tegrating atom and electron detectors and hook up the detectors to the capsule-breaking device. As before, the triggering
of the device will be by pure chance and we can arrange things
so that there is again a 50:50 chance of the cat's being killed

before we open the box one hour later. We switch the electronics on, pet the cat, lock the box, and go away.

We return one hour later and ask ourselves what has occurred. Again, until we open the box we won't know if the cat is alive or dead. Only then will we have complete information. Common sense tells us that, as with the roulette wheel, the cat is either alive or dead, but in the quantum mechanical case is common sense correct? The answer which is given in many textbooks is that common sense has deceived us and the cat is not either alive or dead but both alive and dead at the same time.

Schrödinger's equation is known to mathematicians as a linear equation, and this property of linearity is a cornerstone of the quantum theory. One consequence of linearity is that if there are, for example, two solutions to the Schrödinger equation, then any combination or mixture of these solutions will also satisfy the equation.

The linear property is not held by the other major theory of modern physics. Relativity is an essentially nonlinear theory; it has only single solutions to problems, and paradoxes analogous to Schrödinger's cat could not exist.

In the case of the Schrödinger cat paradox, solutions in which the radioactive isotope has disintegrated or has not are both equally valid, and so are all possible linear combinations of these solutions. That means that two important so-

for the observer there can be only one of two outcomes—the cat is alive or dead. But in the quantum world the many possible solutions exist simultaneously.

A

WORLD OF QUANTUM

TIME = 0 MINUTES

TIME = 35 MINUTES

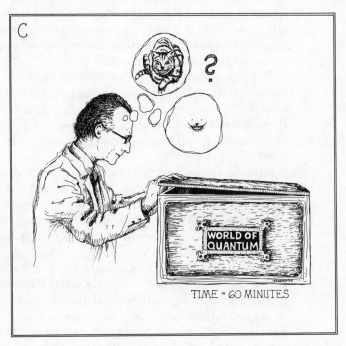

TIME = 60 MINUTES

lutions, one with the live cat and the other with the dead one, are true as well as combinations of cats that are half alive and half dead, cats that are 44 percent dead and 66 percent alive, and so on. Some authorities believe that this means the cat rests in a state of suspended animation inside the box, a clearly ridiculous idea.

What happens when the box is opened? In quantum mechanics this is equivalent to making a measurement and involves the factor that disturbed Einstein—the collapse of the wave function. When the box is opened we learn the fate of the cat, for we see it either alive or dead. Only a single outcome to Schrödinger's equation now applies because only a single wave function has collapsed or been singled out from the mixture or linear combination. It therefore appears that *the act of human measurement has resolved the cat's fate.* How can quantum theory live with such a paradox?

According to Bohr's Copenhagen interpretation, however, the quantum mechanical case is no more mysterious than the roulette wheel situation. In both cases the box lid must be opened to learn the fate of the cat, and this collapse of the wave function is simply an expression of the fact that observation has turned uncertainty into knowledge. The wave function is simply an expression of information and has no objective reality. Talk of a collapse of the wave function or cats in suspended animation comes about because scientists try to press for too much detail.

To some scientists, this explanation has been far from satisfactory. If quantum theory is complete, then the Schrödinger equation must give a complete description of what happens. It must describe the actual mechanism of the collapse of the wave function and give a proper account of the act of observation.

Since cats are not normally both alive and dead, a few scientists have argued that while the Schrödinger equation applies perfectly well to elementary particles, atoms, and molecules it does not apply to living things. At this level of complexity (some physicists think) quantum mechanics breaks down and the Schrödinger equation is not flexible enough to explain the phenomenon of Schrödinger's cat. But where does one draw the line between the micro and macro world, the world of atoms and the classical world of cats? What rules does one apply to tell when that boundary has been reached? Small molecules react together and become incorporated into larger molecules. These, in turn, are taken up to make macromolecules like DNA which are incorporated into the cell. Does the quantum world stop at the atom? The molecule? Where? How many atoms must be grouped together before the collection ceases to be quantum mechanical and becomes "classical"? Eugene Wigner places the ultimate breakdown at the level of the conscious mind, the observer. Wigner is another of those physicists who, present since the early days of quantum theory, has had misgivings about it.

Wigner analyzed the Schrödinger cat paradox and concluded that "quantum mechanics, in its present form, is not applicable to living systems, whose consciousness is a decisive factor."[48] Cats and humans, in Wigner's judgment, have a significant influence on quantum events. If Wigner's analysis is correct, it is quite possible to accept the mixing of states for an inanimate universe, but once conscious beings are added, then wave functions collapse and specific outcomes occur. The conscious mind is firmly placed in the center of the universe, for consciousness now influences events. The observer determines that which is observed.

Other speculation excited by the measurement problem borders on science fiction.

Some physicists, for example, have proposed that a wave-function-collapsing mechanism actually exists in some particular set of cells in the human brain.

Nobel laureate Brian Josephson has attempted to relate quantum theory to a mystical interpretation of the universe similar to that taught by Carlos Castaneda's alter ego Don Juan. According to Josephson, objective reality is produced out of the collective memories of human society while unpredictable, curious events are, to some extent, the manifestation of the individual will.

Others, unwilling to swallow what one physicist calls "quantum solipsism," advocate a solution to the Schrödinger cat paradox which is even stranger than insisting human consciousness determines events.

Hugh Everett III has proposed that the Schrödinger equation is perfectly valid both for atoms and for brains, and that all its possible solutions exist with equal validity. The reason we don't observe live cats, dead cats, and mixtures of the two when we open the box in the Schrödinger experiment is that there are an infinite number of possible universes. Corresponding to each universe will be one of the solutions to the wave function.

According to Everett, a multiple of universes are created

when the lid of the Schrödinger cat apparatus is opened. In many of these universes will be live cats in different states of activity. In one of the universes will be a dead cat.

Less glamorous attempts to solve the measurement problem have included an effort to build a "quantum logic" which casts the laws of micro reality into a logical system that could make sense to our macro minds. Thus far this attempt has met with only limited success.

It is worth repeating that most working physicists don't worry themselves about these questions but get on with their own research using quantum theory.

PROBLEM 3: THE NONLOCALITY OF QUANTUM WHOLENESS — WHERE HERE IS THERE

When Einstein posed the EPR thought experiment (page 71), he assumed that if a correlated pair of particles flies apart an observer looking at one particle, A, cannot instantaneously affect what happens to the distant particle, B. But recent experiments suggest that this is just what happens.

In the 1960s, British physicist J. S. Bell worked out in detail the type of experimental correlation between paired particles that could be expected from quantum theory and from a hidden-variable theory. Actual tests of Bell's calculations have been made in several laboratories. In one of the most dramatic versions, French physicist Alain Aspect split correlated pairs of photons, fired them in opposite directions, and then shifted a polarizing filter in front of one of the detectors while the photons were in midflight. The results showed that the photon twin at B seemed to "know" what had happened to the twin at A. Photon pairs remained very strongly correlated no matter how the polarizer was shifted. This is something like the story about the Corsican brothers, Siamese twins separated at birth, who felt each other's pleasures and pains so that when one was slashed by a foil, far away in another city, his brother felt the cut.

What does this mean? Physicists generally recognize two possibilities. The photon pairs are traveling at the speed of light (photons are light quanta). The first possibility is that for A here to let B there know what is happening a signal must pass between them that is *faster* than light. The second possibility is what Bohr called the "indivisibility of the quantum of action," the wholeness of the quantum experiment. But since A and B are separated by space, distance, *wholeness in this case amounts to asserting that there is no here and there or that here is identical to there.* This gives space and time a property that is called "nonlocality."

Aspect's experiment appears to prove that at the quantum level there are no classical kinds of hidden variables that operate in a localized way like Brownian motion does. It also points to a curious nonlocality of the quantum system and suggests that the space-time of quantum physics is remarkably different from that of Einstein's relativity. Physicists had been alerted to nonlocality by the double-slit experiment and by Bohr's Copenhagen interpretation, but experiments such as Aspect's have caused some theoreticians to view this feature of quantum mechanics as a serious problem and to feel that in spite of Bohr's strictures, the theory must be incomplete.*

A number of scientists have concluded that the normal space-time ordering of events in the large-scale or macro world must not apply to the micro world of quantum systems.

Spurred on by an attempt to unify relativity and quantum theory, Roger Penrose has spent the last decade working on a new form of space-time ordering called "twistors." Penrose says that from our vantage point the electron in the double-slit experiment appears to go through both slits at once. Yet to the curious space-time of the electron, our whole classical everyday world may appear quite different. The slits themselves may in some sense be engulfed by a single electron.

John Wheeler has suggested that below the level of the atom, space-time breaks apart into a foamlike structure. This foam arises in some yet to be determined fashion from what Wheeler calls "pregeometry." Not only space-time but also elementary particles are expressions of this pregeometry. As Wheeler has said, "You and I look at the sky and we see this and that puffy white cloud floating here and here; the clouds

---

* At least one physicist argues that the experiments *do not* show nonlocality. Heinz Pagels claims scientists have been fooled. The apparent correlation is really a correlation taking place in the mind of the experimenters when they put together two sets of essentially random numbers appearing at their detectors. Pagel's refutation (which is complicated) shifts the focus of the problem. It would make the measurement problem discussed previously all the more urgent. Who is this observer who gets fooled by his quantum experiments?

look like the only thing that's important, yet when we go to study them in more detail, we realize that the water vapor in the clouds is a thousand times more tenuous than the air, and that the proper starting point for the description of the sky is not the clouds but the physics of the air. In the same way, the proper starting point for the description of particles is all this activity (foam), all the time and everywhere, throughout space."[19] In an early version of the theory, Wheeler suggested that positive and negative electrons weren't pieces of matter or quantum energy but "wormholes" in the fabric of space-time.

Carl von Weizacker, a former student of Heisenberg's, believes that space-time relationships can be derived from a "tense logic" which arises from the axioms of quantum theory. David Finkelstein has devoted considerable energy to a "space-time code," an abstract mathematical approach to describe processes which lie beneath the fabric of space-time.

Most of these efforts remain vague and fragmentary speculations, or, to use John Wheeler's phrase, "an idea for an idea."

Other physicists have embraced the alternative explanation for the Bell experiments—instantaneous signals or signals traveling faster than light. If such signals existed (there is no evidence for them), they would supply a kind of "hidden variable" needed to explain what is going on beneath the structure of quantum probability. The idea of instantaneous messages has also attracted the attention of those interested in giving psychic phenomena a physical explanation. "Superluminary" signals have been designated the carriers of psychic messages.

If such ideas illustrate anything, it is the confusion that exists when physicists attempt to penetrate deeply into quantum theory—a theory which offers no models and no explanations for reality.

Yet, understandably, the question beckons: How is one to explain this mysterious quantum wholeness?

CRISIS?

Bohr had said in his Copenhagen interpretation that the quantum paradigm was paradoxical. At what point does paradox become confusion? Some scientists are beginning to feel that point has been reached. Are the laws of nature eternal and contradictory? Some commentators, including Kuhn, feel that contemporary physics, now dominated by quantum theory, relativity, and the pursuit of grand unification, is in a paradigm crisis. And not only physics. Though more remotely, signs appear of coming crises in biology and brain science. The work of the looking-glass scientists discussed in this book is an effort to resolve the anomalies and problems of their respective fields. Their theories suggest that incipient crises across the broad front of science may have something in common, something buried in the ancient paradigm that has driven science as a whole for centuries and is just now coming into question.

So while many scientists are busily extending the maps of their established paradigms, while current physicists are skillfully dodging the paradoxes of their theories with great aplomb and seemingly inexhaustible success, other scientists are slipping through the maps into another, as yet dimly perceived reality.

But now, at the borderline of relativity and quantum theory, we notice David Bohm. Musing there, he has seen through the glass to the other side, to a glittering wholeness, like a sea. We notice he has put up his hand to the barrier and like a bright silvery mist it has melted. He has entered a new universe.

2

ZENO'S PATH to DUCK POND

David Bohm's Looking-Glass Map

. . . and jumped lightly down into the looking-glass room. . . . I've cut several slices . . . but they always join on again.

5. From Ducks to Rabbits

As a boy, David Bohm had a vision. Retrospectively, it seems to contain the seed of his theory of the implicate order.

Bohm grew up in Wilkes-Barre, Pennsylvania, the son of a used-furniture dealer. In his early years he liked to climb the hills that surround the town and look down on the streets and houses below. In later life he recalled a peculiar, strong insight he had on one such occasion. He had been thinking about nature and the fact of his own existence when he became overpowered by seeing the lights from the town. The energy from these lights, the young Bohm realized, went out from the town, extending beyond the earth, until it filled the universe itself. Just as his own thoughts seemed to travel without limit, so light energy moves through the universe without end. Nature herself is a web of living energy, each object a mirror made up of strand upon strand of all that is.

Bohm has recently retired from his position as professor of

David Bohm

theoretical physics at the University of London's Birkbeck College. Pale, self-effacing, and apparently reticent, he becomes surprisingly animated and intense when the subject is the nature of reality. At a time of life when most physicists of his stature would have been resting on past laurels, Bohm has given birth to his sweeping "implicate order" theory of the universe. It includes deep insights into such traditionally nonscientific subjects as truth, self-deception, insight, and language—which he has tried to show are as important to understanding the physical world as the classical concepts of momentum and charge. In his theory Bohm has set out not only to devise a new map for a new universe but to create a new understanding of the relationship between maps and terrains. The appearance of this unprecedented approach has fired the imaginations of both scientists and theorists in other fields. It emerges from the soil of the many projects Bohm engaged in throughout his long scientific career.

THE MAKING OF A CARTOGRAPHER

After completing doctoral work with Robert Oppenheimer at the University of Southern California during World War II, David Bohm was offered a position at Princeton University. There he met Albert Einstein. During an intense six months, he and Einstein engaged in lengthy conversations. They discussed the nature of physical theories and the state of quantum theory. They found they agreed that quantum theory had achieved spectacular success in treating the atomic world, but that its claim to be the complete theory of microscopic processes was unacceptable. They considered the problems of extending the theory of relativity into the atomic domain and of formulating a single theory that would account for matter and the forces that act on matter. They did not resolve these problems, of course, but the conversations served as an important stimulus to Bohm's subsequent thinking.

Shortly before he met Einstein, Bohm had decided he needed a better understanding of the subtleties of quantum mechanics and so undertook to write a text on quantum theory. When he finished it in 1951 and gave Einstein a copy, the elder scientist enthusiastically declared he hadn't completely understood the theory until he read Bohm's book.

Pauli was also approving, but Bohr, to whom a complimentary copy was sent, did not return a comment. Bohm's book eventually became a classic text in the field and an inspiration to later generations of physicists. Though it presents quantum theory in an orthodox form, its author stresses features which give the flavor of his own early probing thoughts about the subject:

Quantum concepts imply that the world acts more like a single indivisible unit, in which even the "intrinsic" nature of each part (wave or particle) depends to some degree on its relationship to its surroundings. It is only at the microscopic (or quantum) level, however, that the indivisible unity of the various parts of the world produces significant effects.[4]

After writing the book and lecturing on it, Bohm found the more he knew, the more he "didn't understand what quantum theory was about." There was, he became convinced, some deep confusion in the theory.

Bohm's first research to gain wide scientific recognition was carried out with his student David Pines. Their topic was the collective motion of the astronomical number of electrons that make up a metal. At that time, physicists treated metals as a sea of free electrons locked into a vibrating atomic lattice of nuclei. Many of the measurements scientists obtained when they looked at metals in their laboratories were explained as small excitations of individual electrons interacting in this background sea against the atomic lattice. This approach was an extension of the Newtonian billiard-ball idea of the world. Bohm speculated that something new and completely non-Newtonian was involved. In addition to the haphazard individual fluctuations of electrons, he saw there was a collective motion involving the electron sea as a whole, "an electron plasma."

In such collective modes, the movement of individual electrons might superficially appear to be "random," but the cumulative effect of minute fluctuations in an enormous number (10^{23}) of electrons combined to produce an overall effect. Such collective effects were eventually well established experimentally and called "plasmons." Bohm showed that mathematically the motion of a plasmon reflected the behavior of every electron in the metal. Conversely, each electron implied or concealed the motion of the plasmon as a whole. Bohm's work produced a new and powerful insight into matter, for it showed how orders of collective motion could be concealed or implied in individual explicit movements. Referring later to his research on plasmons, Bohm remarked that he'd often had a distinct impression the electron sea was "alive." Though the traditionally trained physicist in him might be abashed at such a statement, this sense of the aliveness of inanimate matter would one day prove to be a deep insight of Bohm's hypothesis.

In other early work, Bohm discovered that what are called "canonical transformations" of classical physics also suggest motions like those he'd discovered in the plasmons. A canonical transformation is a mathematical reformulation of a familiar law. By putting the old law in a new and subtler form (which is mathematically equivalent to the old form), hidden aspects implied by the old laws can be revealed. In his work on plasmons and theoretical reformulations of established physics, Bohm discovered that deeper orders are hidden in the movements that appear superficially random and that at these deeper levels structures are enfolded and spread out in novel ways.

In his second book, *Causality and Chance in Modern Physics*, published in 1957, Bohm took up another question—causality. Here he argued forcefully that the usual view of causality is far too limited. We ordinarily think of an effect as having only one or a few causes. In fact, the cause for any one thing is everything else. To understand completely the cause of malaria in humans, for example, requires understanding not only the life cycle of the anopheles mosquito, but also evolution, ecology, chemistry, and eventually everything in the universe. Bohm insisted that though for practical purposes many causes are negligible and can be ignored (we needn't consider the whole of evolution in order to make a malaria vaccine), insufficient thought had been given to the implications of the fact that the universe as a whole is a moving causal network.

About this same period, Bohm applied his idea of collective motion to the hidden-variables argument then still raging in quantum theory. His approach was very different from earlier attempts, including Einstein's. Suppose, Bohm proposed, that hidden variables are totally unlike those air molecules that bombard a dust particle in Brownian motion. Suppose they are not only not localized but also not particlelike at all.

Such curious variables would not be excluded by von Neumann's proof or Bohr's refutation of the EPR experiment. Their essence would be the nonlocal correlations eventually demonstrated by the experiments testing the hypothesis of Bell.

The physics community did not react kindly to Bohm's approach. In particular, the followers of the Bohr-Heisenberg Copenhagen interpretation felt that Bohm was trying to turn the clock back to a purely deterministic theory of matter. Even Einstein felt that in his undetectable-hidden-variable analysis Bohm had "got his results too cheap."

Bohm, in reply, said that he had no intention of reintroducing classical determinism. He had simply demonstrated that quantum theory was not the single, exclusive, "complete" explanation for the motions of microscopic matter that Bohr claimed. His own attempt was not supposed to be definitive but merely to open the door for more satisfying theories. Unfortunately for his career, Bohm's message in this argument was profoundly misunderstood. For example, a well-known history of quantum theory by Max Jammer portrayed him as a determinist. Though, as the journal *New Scientist* observed in a recent issue, "Bohm is probably as far from being a determinist as any physicist in the world today," Jammer's book became one of the major sources of information about Bohm's theories. It is perhaps not surprising that the physics community, dominated by the quantum paradigm, came to regard Bohm as a maverick, a brilliant scientist who refused to accept the received wisdom of Copenhagen—someone to be lumped together with Einstein, de Broglie, and other eccentric unbelievers.

In 1965, Bohm produced another book, this time on special relativity. Here he stressed the role of perception in science. He noted that psychologists have shown that our perceptual apparatus (eyes, ears, etc.) abstract relatively unchanging or "invariant" features from the environment to create mental maps. A map of a highway doesn't show the pavement with its changing potholes and patches or the shrubbery along the way, but only abstracts (pulls out) certain features such as the curves and direction of the roadway. The mechanisms of memory and perception do the same thing. They store and react to relatively "invariant" aspects. Once a mental map is formed, it then conditions further perception. For example, someone walking down the road might see a small object moving across the pavement ahead of him that he recognizes is a chipmunk. He clearly "sees" this chipmunk until he gets closer and realizes it's only a dead leaf being moved by the wind. According to psychologists, the very real experience of the chipmunk resulted from his having applied a complex set of abstract features in his memory to the object in front of him. What he "saw" was not "out there," nor was it entirely an illusion; it was some relationship between the old maps he applied and the movement of the object which later turned out to be a leaf. Like Kuhn, Bohm stressed that scientific theories are also maps that guide how and what things are seen.

He insisted, however, that this did not imply solipsism— the world as the perceiver's illusion. It implied something more profound. "The very fact that our vision of the world can be falsified as a result of further movement, observation, probing, etc., implies that there is more in the world than what we have perceived and known."[6]

Bohm concluded from these reflections that the purpose of scientific research is not the accumulation of knowledge. All theories eventually are falsified. The meaning of scientific research is that it is an act of perception, a continual process of consciousness and nature.

This insight appeared to mark a turning point in Bohm's thinking. Now the various underlying strands of his years of research and musings finally converged and became apparent in his notion of the implicate order.

SLICING THE CAKE

In light of Bohm's history, it shouldn't be surprising to learn that his hypothesis completely opposes a premise embedded in the thinking of scientists since Aristotle. The premise he attacks is that nature can be analyzed into parts. In its place, he proposes to consider the universe an undivided whole. Many other theories, of course, have referred to the idea of wholeness, including, as we've seen, both quantum theory and relativity. But, in practice, science has persistently dealt with nature as if the parts and bits pressed under microscopes and accelerated in particle chambers were real. Wholeness, Bohm points out, is one of those ideas almost everyone pays lip service to but virtually no one takes seriously enough to find out what it means. For to take undivided wholeness seriously is to make an incredible journey, abandoning everything that is familiar and comfortable. It is as weird as the idea of slipping through to the other side of a mirror.

In his writings and talks David Bohm painstakingly works to bring attention to the subtle difficulties involved in understanding the difference between the fragmentary approach that has so long dominated science and an approach that assumes wholeness. His argument goes as follows:

The fifth-century Greek philosopher Democritus proposed that the world was made of atoms moving in a void. He imagined all large-scale objects were the results of different arrangements of atoms. Perhaps initially Democritus' view was an "insight" into wholeness because it allowed human beings to understand how the whole of the huge variety of objects in the universe could have an underlying unity at the atomic level. For Bohm an "insight" isn't a fixed truth but an act or angle of perception. An example of insight might be learning to sink a hook shot or ski a difficult slope. The insight finds a relationship among moving elements which makes sense. In this case, the "sense" is hitting the basket or making it to the bottom of the mountain without falling. The knowledge involved in this process is not fixed or final, because it depends on, and must adjust to, constantly changing circumstances. If a basketball player tried to anayze and turn into fixed knowledge all the steps in making a hook shot, he probably couldn't make one. Insights involve recognition that there are aspects going on that are beyond analysis. Bohm wants to construct a theory of physics that will provide "insight" and not just knowledge—because he believes that failing to do so has devastating consequences for wholeness. These consequences, he believes, can be illustrated by looking at the fate of Democritus' insight about atoms.

Democritus' insight lost its elasticity and hardened into a fixed truth as science evolved. As the atomic idea became more and more articulated, instead of calling attention to the undivided nature of existence, it became the basis for assuming reality is made up of fragments. Democritus pointed his finger at an underlying unity, but after a while people stopped looking where he was pointing and began studying his finger. Once the shift from insight into "knowledge" occurred, everything was thought to be composed of parts—parts existing independently and lying outside of each other, connected by external relationships. Eventually this fragmentary point of view became incredibly complicated.

Consider as an example what has happened in the medical and biological fields—where one might have expected the idea of parts to have had little appeal. After all, isn't the body one organic whole? In principle, most physicians and biologists would agree that it is. But in practice, the medical profession is divided into specialties, each treating a different part or system of the body. The heart is, on one hand, inextricable from the whole, but on the other hand we can replace this part with another through transplantation. Bohm emphasizes that science is literally pervaded with the sense and experience that while things may be tightly interconnected—like the heart is with the bloodstream, liver, bone marrow, and muscles—they are nonetheless fundamentally separate and analyzable.*

And it's not only science. A moment's reflection will reveal how virtually every aspect of human thought relies on the notion of parts. An instant's reflection reveals the magnitude and pervasiveness of this subtle, tyrannical and apparently inescapable assumption.

WHOLE GAPS IN THE FRAGMENTS

Now, however, Bohm thinks, the assumption is being seriously challenged—by the very human pursuit that has taken the most systematically atomist approach—Western science. That approach has led inadvertently, or perhaps inevitably, to an undeniable revelation. Though most scientists continue to practice their profession as if the towering two-thousand-year-old atomistic, mechanistic paradigm were still unquestionably on the rise, Bohm believes it is already trembling at its foundations. The seeds of its collapse are buried in some generally ignored but very deep implications of quantum theory and relativity.

---

* Part 4 describes a looking-glass theory which suggests that transplants are possible precisely because of the holistic nature of the body, not because the body is a collection of interchangeable parts.

Quantum theory has shown that the idea of a separate atomic particle (the ultimate "part") cannot be consistently maintained. At the quantum level, two entities far apart from each other affect each other, showing that they are united non-locally and noncausally. In contexts such as the double-slit experiment a particle loses its definition in time and space. It spreads out and is like an ill-defined cloud whose shape depends on the whole environment, including the instrument that is observing it. Bohr had said that even this ill-defined cloud is not a separate entity; the total experimental situation is an undivided whole. The quantum discovery means that the world view in which an observer and what he observes are separate "parts" of the universe also cannot be consistently maintained. Instead observer and observed are found to be somehow merging.

For its "part," relativity shows the various contractions of objects moving relative to each other. Because of these contractions and the need to account for them, scientists are forced to abandon the idea of a rigid body. Without a rigid body, the idea of separate particles becomes meaningless. In their place emerge "world tubes," each representing the infinitely complex unanalyzable continuous process of a structure in flowing movement, like a vortex in a stream. There is, of course, no sharp division between a vortex (world tube) and the stream. Einstein's general theory of relativity and his attempt to find a unified field theory extended his relativistic idea that the universe is one unbroken whole.

As we mentioned, physicists have put considerable effort into trying to reconcile the differences between quantum theory and relativity. Bohm asserts that this reconciliation can take place only if certain aspects of each theory are dropped. The two theories agree that the universe is whole, yet each harbors elements which contradict the holistic premise. It is precisely these fragmentary elements, Bohm believes, which have prevented the theories from being united.

Relativity, for example, retains fragmentation through its emphasis on "the signal." Special relativity presumes it is possible to send a signal, at the fixed speed of light, from one world tube to another. The signal could be a radio transmission sent from the world tube of earth to the world tube of a rocket ship traveling near the speed of light. The idea of a signal connecting world tubes occupies such an important place in the theory that some textbooks on relativity base much of their treatment on the exchange of signals. Space-time frameworks are said to be constructed by means of accurate clocks and light signals.

But the idea of a signal implies that some separate "thing"—light energy or a radio broadcast—is transmitted from one independent part of a system to another. The idea of autonomous regions of space connected by separate information carried on a signal becomes exceedingly unclear in a context insisting on undivided wholeness.

In quantum mechanics the fragmentary snake in the holistic garden is the wave function, which physicists call "the quantum state of the system." The wave function is assumed by some physicists to be objective. That's why Everett could find such surprising acceptance among scientists for his multiple-universe theory in which all these possible Schrödinger cats really exist. Bohm says this means that when physicists refer to "the quantum state of the system" they are actually thinking of it as a separate entity despite the fact that quantum theory implies that talking about separate entities has no meaning. The same criticism holds for the idea of interacting "fields." In the grand unification theory, quantum fields are conceived of as independently defined at each point of space and time.

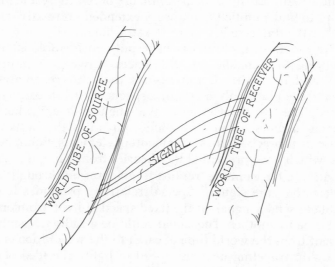

Bohm's analysis of the hidden fragmentary features of quantum theory and relativity raises, of course, a particularly squirmy question: If there are no separate and independent parts to the universe, how can scientists (or anyone else, for that matter) talk about anything at all without falling hopelessly into contradictions? Language, even mathematical language, proceeds by analyzing things, splitting them into cause and effect. We must abstract elements in order to make maps. If we don't have maps, we won't be able to know where we are or get anywhere. Bohm is excruciatingly aware of this problem. Map we must and talk we must, he agrees. But he insists that we should keep in mind that our maps and words are never about absolute "things." At best we communicate an insight, at worst, an illusion.

Thus for Bohm, as for Kuhn, scientific theory is a perspective, a point of view which makes some mysterious contact with nature. Such contact exists in a limited way but ceases to give a useful picture if extended too far. For example, classical physics, which was developed before the invention of the kinds of instruments that allowed scientists to look into the atom, gives a view of the universe that corresponds to a great many everyday sensory experiences, at least for our senses as they are currently conditioned. It works for that dimension. When physicists extended our sensory apparatus by means of spectrography and particle accelerators, we entered another dimension, or deeper level of the one we're in. Here the classical map became confused, and after some struggle scientists eventually had to see the classical view as only an approximation. But as Kuhn has shown, new approximations, new theories, won't bring us any closer to the absolute truth. Bohm believes that each time we come to the place where our old theory ceases to give meaningful answers we have, in effect, discovered the universe is undivided and whole, that it extends always beyond (or deeper) than any map, equation, definition, or theory.

Philosopher Martin Heidegger once offered an analogy to describe the wholeness of truth. Heidegger compared truth to a drinking glass: As you turn the glass in order to see one aspect, you necessarily have to conceal another aspect. You can never see the whole glass although it's all there in whatever aspect you do see.

For Bohm an important image of wholeness is the vortex in a river. From a distance one can clearly see the turbulent water of the whirlpool and the slowly flowing river. These appear to be two separate "things," but approaching closer one notices it is impossible to say where the whirlpool ends and the river begins—such analysis into separate and distinct parts flounders. The whirlpool is not a separate thing at all but an aspect of the whole.

By emphasizing wholeness Bohm does not propose, however, to eradicate distinctions and differences. He cautions us also not to think of turning the universe into one giant undifferentiated, mystical blob, a "Bohm monster." Indeed, he argues that unless we understand the subtleties of wholeness, we will not only divide what can't be divided, *we'll try to unite what can't be united*. Real differences and similarities will become hopelessly mixed up. It's perfectly reasonable to say that two vortexes in a stream are aspects of the whole stream, but it is entirely *unreasonable* to say they are both the same vortex. Without understanding wholeness we will hopelessly confound the relationships between parts and whole. Bohm outlines this confusion as follows:

This can be seen especially clearly in terms of groupings of people in society (political, economic, religious, etc.). The very act of forming such a group tends to create a sense of division and separation of the members from the rest of the world but, because the members are really connected with the whole, this cannot work. Each member has in fact a somewhat different connection, and sooner or later this shows itself as a difference between him and other members of the group. Whenever men divide themselves from the whole of society and attempt to unite by identification within a group, it is clear that the group must eventually develop internal strife, which leads to a breakdown of its unity. . . . True unity . . . between man and nature, as well as between man and man, can arise only in a form of action that does not attempt to fragment the whole of reality.[7]

The definition of a tree as a thing or part of nature composed of roots, trunk, limbs, and leaves interchanging with the environment is useful if we want to cut trees down or plant them. In a larger context, however, this idea may be detrimental. The tree is not a part. It is impossible to say at just what point a molecule of carbon dioxide crossing the cell membrane into a leaf stops being air and becomes the tree. The tree threads out into the whole environment and eventually the whole universe. If this fact is ignored and forests are cut down, consequences will arise which affect the whole ecology. Human misapprehension about parts and whole can be not only confusing but also dangerous.

For Bohm this kind of part-whole jumble is sown when scientists convince themselves that a theory which makes sense out of a fact is really separate from the fact that it investigates. Bohm believes, as does Kuhn, that facts are not self-contained realities; they are "made" by theories. Facts are abstractions of certain aspects out of the unbroken flow. Scientists' theories or instruments (which are mechanical extensions of theories) give these aspects their "shape." Quantum physicists know, for example, that one kind of experimental setup will produce quanta as particles; another kind of apparatus will make quantum waves.

In a larger sense, facts are made by the way a scientific theory orders the universe. In classical physics, facts were made by the theoretical order of planetary motion measured by position and time. In quantum theory, facts are made by an order which includes energy levels, quantum numbers, symmetry groups, and measurements in terms of scattering, cross sections, charges, and masses of particles. During paradigm shifts, changes in theoretical order lead to new ways of doing experiments and to making new facts.

Bohm believes, therefore, that most contemporary quantum physicists are involved in a self-deception. At one level, Bohr's Copenhagen interpretation forces them to acknowledge that the facts they discover like photons and mesons and other particles are abstractions somehow related to their experimental devices. At another level, however, they treat these facts in every respect as if they existed independently of the devices and theories. Bohm isn't surprised that scientists have found so many particles, that these particles often dissolve back into each other, or that an ultimate grandfather particle has proved elusive.

If this sounds the least bit mystical, it isn't. Though wholeness is a traditionally mystical idea, Bohm himself is a thoroughgoing realist. Anyone reading or listening to him recognizes immediately that he is a man dedicated to achieving no-nonsense solutions to the problems of his profession. It's just that, as he sees it, the problems now require a new order of insight into nature—an insight which is clear about wholeness. The challenge of achieving such an insight and describing such an order is obviously monumental. If an order of wholeness is to have even a chance of being taken seriously by science it will have to (1) go well beyond the statement "All is one," and give wholeness an understandable and somehow concrete form, something which even Bohr and Einstein failed to do, and (2) give a good account of why the world of our everyday experience seems so definitely composed of parts. On both of these counts Bohm, at least conceptually, succeeds in describing his new order.

Before we see how he does it, we need to be clear about just what he means by this word "order."

DESCRIBING THE ORDER OF ORDER

There are all kinds of order. There's the order of the whole numbers in mathematics: 1,2,3,4, . . . where each number is thought of as an equal "distance" from the next, like the inch marks on a ruler. There are more complicated orders like the "powers" in mathematics, 2^2, 2^3, 2^4, etc., which involve more complex relationships. There's also the order of a Bach cantata and the order implied in the transformation of an acorn seed into an oak. Bohm says a general way of perceiving what is meant by order (not a definition, mind you) is to say that order means "to give attention to similar differences and different similarities." For example, suppose we plot a "curve" made of straight lines going in different directions:

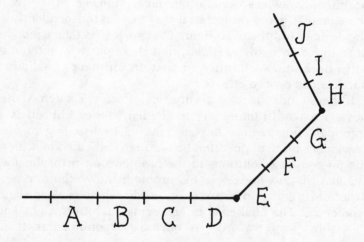

The points on the line ABCD are obviously similar and we can readily see their order. EFG is another order, and HIJ is a third. So we have three degrees of similarity. But the lines ABCD, EFG, and HIJ are also different from each other, so first we see the differences of their similarities and then go on to see the similarity of their differences. Got that? If you do, then you can undersand that it is possible in principle to describe incredibly complex degrees of order by giving attention to the arrangements of similarities and differences.

The opposite of order is popularly thought to be disorder or randomness. The movement of dust motes dancing in a sunbeam (Brownian motion) might be called disordered or random because scientists can't describe the similarities and differences with a finite number of steps—there are just too many. Nevertheless, Bohm says it would be inaccurate to say Brownian motion is disordered. Instead, he wants to say it is a very special kind of order—an order of "indefinitely high degree." In fact, one of the major changes Bohm makes as he moves from ducks of parts to rabbits of wholeness is to replace the idea of "random" order with the idea of a hierarchy of order. Intuitively, one can see his reasons. Remember the double-slit experiment? When scientists fired particles at the slits one at a time, they couldn't tell which slit any particular electron would go through or where it would land. So far as the experimenters were concerned, the electron movement was random. Nevertheless, when the experiment was completed, the electrons didn't show a random blur on the detection screen. In experiment after experiment, the particles formed a beautiful and entirely predictable wave on the photographic plate. The "randomness" somehow contained a very high degree of order. So for Bohm there is no randomness, only various degrees of order. Logic also requires Bohm to take this stance: If one assumes that the universe is whole and unbroken, it wouldn't make sense to say that some parts of it are ordered and other parts random, for such an idea would contradict Bohm's premise by implying the universe has ordered and unordered "parts."

In classical physics the idea of order was the so-called Cartesian grid, with time on one axis and distance or space on the other. An event could be plotted (its order revealed) by finding the distance it had traveled in a certain time. This Cartesian order has permeated all of physics and survived each revolution in the paradigms of science. The Cartesian order is essentially fragmentary. A line is constructed from points, a surface from lines. Space is linear, continuous, and infinitely divisible. Time is also linear. The equivalent of the point in continuous space is an "instant" in continuous time.

All qualities of interest to the modern physicist are expressed as functions using Cartesian coordinates. Although the space-time order of relativity is no longer absolute, its Cartesian description has been retained. In quantum theory the idea of the path has to be abandoned; a particle cannot even be talked about as having existence as it jumps from one state to the next. Nevertheless, embedded in quantum mathematics are coordinates with their notion of continuity, paths, and infinite divisibility.

Bohm believes the Cartesian order is no longer appropriate to twentieth-century physics. A new order of description is called for. Relativity relies in part for its order on the notion of a signal, quantum theory on the notion of "the quantum state of the system" or wave function. To achieve a new order Bohm proposes dropping the basic role of the signal and that of the quantum state as old and fragmentary notions tied to the Cartesian system. Doing this will allow quantum theory and relativity to harmonize. Though this sounds simple enough, it is, as he says, "no small thing" to let go of these features, because they are the foundation blocks of an analytical view of reality which has held sway for thousands of years. It is a shift from ducks to rabbits which will require a radically different kind of order—an order which divides itself only as it is passed around.

6. Unwinding Reality's Threads

To illustrate what he calls his "new order of fact," Bohm offers three analogies or models which provide three looking-glass entryways into the new science of undivided wholeness.

MODEL A: THE HOLOGRAM

The first analogy involves photography.

Bohm believes the camera lens is a good example of the close relationship between instruments and theory. A camera lens forms an image of something. Let's use for an example the family at an outdoor barbecue. When the snapshots come back from processing, every point on the little glossy card in your hand corresponds to a region in the tableau that was taking place the moment the shutter was snapped. Of course, the photo is really only an abstraction, a mapping of certain aspects of three-dimensional reality onto a two-dimensional form. Anthropologists have reported that aboriginal people shown snapshots of themselves usually can't see anything but a swirl of abstract colors and shapes. They don't know how to read that kind of map.

Nevertheless, the lens and its ability to abstract features of a scene so that a region on the lens corresponds to a region in the scene make it a powerful model of analysis into parts. Bohm argues the lens allowed scientists to look at objects in such great and part-icular detail that they were encouraged to think that if only they could find lenses powerful enough they could see the parts of everything right down to the electron. Of course, as we've seen, Heisenberg put an end to that belief.

Bohm asks: Is there an instrument that could help gain insight into wholeness the way the lens encouraged insight into analysis and parts? He answers that there is—the hologram.

In Part 5 when we consider Karl Pribram's holographic theory of consciousness, we'll discuss holograms in more detail. For now we need to concern ourselves with only certain aspects. A hologram is a kind of photograph usually made by shining laser light through a half-silvered mirror. Some of the light is reflected by the mirror onto the object or scene being photographed and then reflected back onto a photographic plate. The rest of the light goes directly through the mirror onto the plate (see illustration on pp. 250–51). When the two beams unite at the plate they interfere with each other to produce a pattern. The pattern looks nothing at all like the scene being recorded. It has something like the appearance of a pond into which someone has thrown a handful of pebbles with many, many concentric wave patterns intersecting.

Most people who have seen a holographic image—which is

The holographic plate records, coded in its interference patterns of concentric rings, a three-dimensional image of the rabbit. The interference patterns have been produced by light bouncing off different features of the object. The patterns encode those features. By shining a laser beam through the plate, the encoding can be retrieved so the rabbit seems to appear in space. This can also be done by shining a beam through only a piece of the plate!

produced by directing a laser beam through a holographic plate on which a scene has been recorded—experience the eerie sensation that they are looking at a real three-dimensional object. It is possible to walk around the holographic projection and see it from different perspectives just as with a real object. Only by reaching out to it does one discover there is nothing there. High-powered microscopes placed on a holographic image of a drop of pond water can disclose the same microorganisms that were in the original drop (though in frozen form). But there is a still more curious feature.

If a photographer tore off a piece of the negative of our barbecue image and made a print from it, the print would obviously contain only a piece of the original picture, say Dad's arm turning the rotisserie. If a holographer tears off a piece of the holographic negative and shines a laser beam through it, he gets not a "part" but the *whole* image (though lacking crispness). This shows there is no one-to-one correspondence between regions (or parts) of the original scene and regions on the holographic plate the way there is on a photographic negative produced by a lens. The entire scene has been recorded everywhere on the holographic plate, so each and every "part" of the plate reflects the whole. For Bohm, the hologram is a forceful analogy for the whole and undivided order of the universe.

What is happening on the holographic plate that produces this effect where all "parts" contain the whole? As Bohm sees it, what happens at the plate is simply a momentary, frozen version of what is occurring on an infinitely vaster scale in each region of space all over the universe.

Light and other waves of electromagnetic energy travel infinitely, constantly interfering with each other as they reflect off matter. These interference patterns are endlessly evolving "encodings" of these reflections of matter. Thus the flowing, changing interference patterns traveling through space contain incalculable amounts of information about the objects they've encountered. Specifically, they contain information about the various orders contained in objects—orders involving such features as an object's geometric forms, the relationship between its insides and outsides, intersections and separations. The hologram—which uses an especially "simple" brand of energy to create the interference patterns—shows some of the encoding potential of interference patterns in general.

Now, a turn of the screw. Remember that matter is also waves. *Therefore the very matter of objects is itself composed of interference patterns which interfere with the patterns of energy.* What emerges is a picture of an encoding pattern of matter and energy spreading ceaselessly throughout the universe—each region of space, no matter how small (all the way down to the single photon, which is also a wave or "wave packet"), containing—as does each region of the holographic plate—the pattern of the whole, including all the past and with implications for all the future. Each region will carry this encoding of the whole somewhat differently, as in fact different "parts" of a holographic plate will each give the whole picture but with slightly different limitations on the number of perspectives from which it can be seen.

It is a breathtaking view, an infinite holographic universe where each region is a distinct perspective, yet each contains all.

But lest it seem too far from our everyday experience, Bohm says, recalling perhaps his youthful vision on the hill above his hometown:

Consider, for example, how on looking at the night sky, we are able to discern structures covering immense stretches of space and time, which are in some sense contained in the movements of light in the tiny space encompassed by the eye (and also how instruments, such as optical and radio telescopes, can discern more and more of this totality, contained in each region of space).[7]

With the holograph analogy we arrive at principle number one of an unbroken, holistically ordered universe: *Everything mirrors everything else; the universe is a looking-glass.* The coffee cup in your hand, your hand, the patch of light on the kitchen wall, all the features which we identify as parts carry an enfolding of the whole in their interference patterns. Bohm's vision recalls the poet William Blake's famous lines:

To See a World in a grain of Sand,
And Heaven in a Wild Flower,
Hold Infinity in the palm of your hand,
And Eternity in an hour.

MODEL B: THE IMPLICATE DYE DROP

Bohm's second model to describe the properties of his implicate order is his favorite and captures looking-glass principle number two: *Wholeness is flowing movement.* The analogy goes like this:

A drop of dye is placed in a viscous liquid such as glycerine which is encased between two glass cylinders, one inside the other. The outer cylinder is rotated slowly. As this is done, the dye drop threads out into the liquid. After a number of turns of the cyclinder, the dye appears to have totally disappeared. In conventional terms the distribution of the dye would be said now to be random, with the initially ordered state (that is, the drop clearly, explicitly at one place in the liquid) passing into one of higher entropy (disorder)—information has been lost and explicit order destroyed.

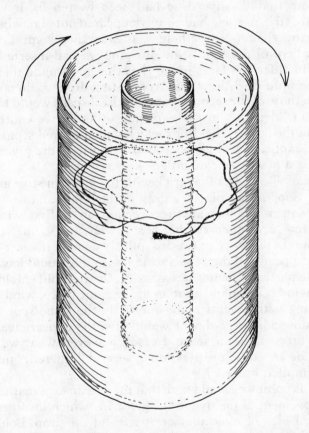

GLYCERINE DYE EXPERIMENT

Bohm often surprises listeners by asking what would happen if the cylinder were rotated backward the same number of turns it was rotated forward. Surprisingly, the drop would reconstitute itself! We would discover that the apparently random state had not been one of disorder at all but of a hidden or "implicit" order. Bohm says this is the kind of order that pervades the universe. He calls this hidden order which is spread throughout the whole an "implicate" order (from the word "implicit"). When it evolves into a form which can be seen (like the drop of dye), it becomes "explicate." Implicate-explicate orders are different expressions of the same order—two sides of one coin.

In the glycerine-dye experiment, if we didn't know beforehand that the dye drop had been wound in, it would seem this dye particle had simply appeared out of nowhere in glycerine space. Nevertheless, it was there, spread out everywhere in the fluid. Now imagine we had inserted two drops in different places in the glycerine and wound them in. In this implicate order, threads of the first drop would now lie next to those of the second drop. But the threads would have a different "destiny" in the explicate order. If we continued turning the device, the drops which had appeared separately in the explicate would be wound again into the glycerine, forming a new implicate order.

We can also do another trick with the glycerine-dye experiment. Suppose we placed a drop in the liquid and wound it in a few turns and then another drop next to the first, winding that a few turns, then another next to that, and so on. Now, if we wound the device backward quickly enough the last drop would appear and then the next to last and it would look like a single particle moving across space. Or we could enfold one drop, wind it in, enfold another at the same place, wind that, following with several more in exactly the same spot. This time, unwinding the device would give the appearance of a single drop stable in space. In fact, however, what we saw would be an explicit expression of several "different" underlying implicit orders.

At this point we should note that there is an explanation for the movement of the dye in the glycerine which involves the classical idea of molecules or "parts" of the drop. Bohm is using the cylinder device only as an analogy or model. In the subatomic and supergalactic universe he is talking about, there are no such parts. Instead, apparent "parts" are seen as world tubes into which one can go deeper and deeper as one can go deeper into the implicate order of heavens by using ever more powerful telescopes. In the past, physicists have been concerned to explain apparently separate particles and the tracks of particles moving across space, plotting them on Cartesian coordinates. Bohm uses the glycerine-dye experiment to demonstrate that it is possible to focus on a completely different kind of order. What he proposes is to focus now on implicate orders which are not made up of parts but are orders in which "things" enfold one another. These implicate orders provide the ground (though, as we'll see, not necessarily the ultimate ground) for the explicate orders we call particles and planets.

Bohm calls his looking-glass physics the physics of "holomovement." A radio wave can "carry" enfolded in its movement various orders which can be *un*folded by the electronic circuity of a TV into a two-dimensional moving image. With the hologram, the movement of interference patterns of coherent (laser) light enfolds a much subtler range of structures and orders. When these are recorded on a plate and retrieved by a laser beam, the viewer sees three-dimensional scenes from many points of view. In a similar, but unthinkably vaster, way, the whole movement or "holomovement" of the universe carries the implicate order and allows us to see and experience our four-dimensional space-time world.

The mind can grasp the principles of movement in the glycerine carrying the dye drop, or movement which carries the order in a television wave. It can even grasp movements of laser light carrying the complex orders of a hologram. But what kind of *movement* is it which carries the implicate order of the universe?

The concept of movement is not an easy one. The early Greek philosopher Zeno of Elea first demonstrated the serious problems with it when he presented his famous paradoxes of motion. Zeno showed that the kind of apparently continuous motion we see every day is impossible. Zeno argued that in order for an object to travel from A to B it would first have to go half the distance to A and then half of that half and then half of that half and so on. Visualize the journey as a line which you can keep dividing up infinitely into smaller and smaller distances beween points. No matter how small you make the divisions, there is still the problem of how to get from one point to the next without some discontinuous (what theorists later called quantum) jump.

A century later, Aristotle offered a resolution to Zeno's paradoxes. In effect, he visualized the points as overlapping one another. With the Cartesian order of continuous functions and Newton's and Leibniz's invention of the integral calculus, mathematicians discovered a trick for summing up an infi-

Zeno said the duck can't get to the pond because the distance between two points can always still be divided, so there'll always be a step to go. The quantum rabbit solves the problem by jumping discontinuously from one point to the next. But nobody knows how she does it.

nite number of steps without getting an infinite answer, so Zeno's paradoxes appeared trivial. Yet with the advent of quantum theory, the notion of a smooth, continuous path suddenly disappeared and Zeno's paradoxes were viable again. How do things get from one point to the next?

Bohm says both the continuous and discontinuous ideas about movement make a fundamental error. They view movement abstractly and aren't faithful to the way movement is actually experienced. The traditional approaches to movement view it as taking place over time. For example, a person walks from one side of the room to the other. This takes a certain time. Each step is like a tick on the clock or a point on a line. The movement is viewed by scientists and the rest of us conceptually as if the whole trip were present all at once. Movement is described as both its beginning and its end. If someone asked the person what his movement was he'd say, "I went from here to there," from one point in the room to another. But was this really his *experience* of movement while he was moving?

No doubt the actual experience of the traveler crossing the

room was that at each moment his movement was present to him and earlier moments of movement were nonexistent; future moments of movement were equally nonexistent. Trying to take the past moment of movement and the present moment as if they were at the same time, the way Aristotle did when he overlapped points, is remote from experience. What in fact does it mean to take past and present at the same time? The idea seems inherently confusing. Views of movement which rely on the concept of points on a line are, Bohm says, useful for solving certain kinds of problems but remain, at bottom, exceedingly abstract. And they don't help us at all when it comes to movement in the quantum world.

Movement in terms of the implicate order is closer to our intuitive sense of movement. Think of the series of dye drops enfolded next to each other so that as they unwound quickly they appeared to be a single particle moving across the glycerine. In its implicate form we recognize that this "particle" of dye was in fact not an "object" at all but a series of interpenetrating elements (or strands) in different degrees of enfoldment. From the implicate point of view, we learn the movement of the "particle" was not that of a object traveling from one point to another across space and time; the movement was different degrees of unfoldment all present *at the same time.* So instead of describing movement as one point related to another, in the implicate order movement is described as *one form of present (one degree of enfoldment) related to another form of present (a different degree of enfoldment). All these different "presents" are unfolding together at any moment.* There is no abstract Aristotelian or Zenoian line full of points showing past and future. Bohm's approach obviously also has major impact on the concept of time, which we will move to in a moment.

Making his case from another direction, Bohm points out that Einstein's relativity described movement as a signal which cannot exceed the speed of light. But Einstein himself had described another kind of movement which implies velocities that do exceed the speed of light—Brownian motion. Physicists say that with Brownian motion the particles being kicked around by atoms and molecules attain "instantaneous speeds." However, Brownian motion cannot carry a signal. A signal requires an ordered modulation such as takes place in a radio wave. An ordered modulation doesn't lose or mix up the information in a way that would prevent it from being sorted out at the other end, say in a TV set. Brownian motion (which is discontinuous motion) is usually considered random. But Brownian motion, Bohm claims, isn't really disordered; it's just a very "high degree" of order. This order is the relationship of various different "ensembles" of order unfolding together at the same time, something like the threads of a huge number of dye drops affecting each other in the implicate space of a glycerine cylinder. The order of Brownian motion is so high that for practical purposes any fixed information carried in it would become altered in the sending—an ensemble of information would get mixed up by other ensembles.

Bohm proposes to use Brownian motion to "rewrite" relativity theory. He wants to say that the laws of relativity apply and are appropriate when the "average" speed of Brownian motion does not exceed the speed of light. When that is the case, we are looking at an explicate order. In this explicate order we can talk about such "things" as signals. Beyond that explicate order are deeper orders—Brownian motions of very, very high degrees. These deeper orders are implicate orders, enfolded and unfolding orders which seem to our senses random and discontinous. By saying the laws of relativity are laws of an "average," not absolute laws, Bohm harmonizes relativity with quantum theory, which is itself a law of averages (probability). So beneath the events described by quantum theory and relativity, Bohm says, are deeper movements and orders which must be described by deeper laws—laws of the implicate order.

Referring to the observations of child psychologist Jean Piaget, Bohm offers that our earliest experience with the world is implicate—it is flowing Brownian movement, movement of an enfolded-unfolding continuous "present." It is only through learning that we acquire the shared maps of the world which include the Cartesian point-broken time line, as well as a sense of stable continuous separate things and parts. These maps constitute and describe what Bohm calls the explicate order. Language is a highly advanced explicate map, since it depicts the world as separate stable fragments and parts organized into knowledge. In its written form, this map can be kept unchanged and passed on to future individuals. Civilization is thus equated with higher and higher degrees of explication. Perhaps for this reason, cultures which do not possess writing seem closer to an implicate perception of things flowing into each other than do highly "civilized" cultures such as ours.

One might be tempted to wonder at this point: If everything is in fact flowing movement, how can we possibly ever have made those maps, which, after all, do give us the ability to perform quite remarkable feats with reality like going to the moon and liberating vast amounts of atomic energy? Surely parts and pieces can't be all illusion.

We can illustrate Bohm's answer to this point with another analogy. Scientists who study the oceans have discovered that they are not just one boiling mass of liquid pulled on by lunar gravity. Most people are familiar with currents like the Gulf Stream and the Humboldt. There are also several layers of "separate" currents thousands of feet deep which extend around the globe. Scientists have found that these currents remain stable, with different temperatures and different directions from the currents above and below. So though the ocean cannot be treated as made up of parts or things and there can be no real separation between one layer of water and another, there is nevertheless a kind of relative separation or, to use Bohm's phrase, a "relative autonomy" to these currents. It's easy to see that this relative autonomy derives from the whole motion of water and in turn any wave or swell or undersea whirlpool anywhere in the ocean "implicates" that whole.

In order to describe such "currents" as stones and particles which appear in our everyday explicate reality, Bohm uses a rather formidable-sounding phrase. He says such things are "relatively autonomous subtotalities." Though the phrase may be awkward, it is precise. It indicates that things like stones and particles and undersea currents may be considered separate but only relatively so and that they are whole in themselves but with a wholeness that derives from the larger whole (though the idea of greater and lesser "wholes" is really only an abstraction). In order to make this a bit easier to read, from here on we'll sometimes abbreviate the phrase "relatively autonomous subtotalities" to the single word "subtotals." Bear in mind that beneath this word is Bohm's whole subtle idea.

Bohm wants to make us aware that what makes subtotals stable is not their separateness but the movement of the whole. So instead of speaking of separate things (such as particles) in interaction, Bohm speaks about relatively autonomous movements which are limited and made stable by other relatively autonomous movements, the way the movement of the Gulf Stream is limited by the movements of other currents in the North Atlantic. The laws which apply to any subtotal (laws such as those which describe the movements of billiard balls and planets in classical physics or such as those describing particles and fields in quantum physics) are always limited and made stable by a larger law of the whole. Bohm calls this law "holonomy."

From the point of view of holonomy, elements which are going to appear together in explicate form (like the threads of the dye drop before the drop itself becomes visible) constitute an ensemble, a subtotal bound together by "the force of overall necessity."

The ideas of holonomy, ensembles, and subtotals led Bohm to propose a new way of approaching physics. For thousands of years science has concentrated only on the explicate orders of the universe. He wants scientists to realize now that beneath each explicate order lie implicate orders, as the threads of the dye drop in the glycerine background lie beneath the dye drop itself, or as the phenomena of quantum physics lie beneath those of classical physics. Physicists can now set themselves to investigate the relationship between explicate subtotals and the implicate ensembles which give rise to them. The object of this study will be to formulate the laws of necessity which bring particular sets of intermingling ensembles into explicate form. At the same time, physicists will acknowledge up front that there can never be a fundamental overall law of the whole, a permanently defined holonomy. There can be no absolute and final theory for all of physics. Instead, physics should view itself as a process with nature. As each implicate subtotal layer is explored, this will become more and more explicate to our understanding; for example, we have seen how the phenomena of quantum physics were at first a totally mysterious implicate layer beneath the classical and then became more and more explicate as Bohr, Heisenberg, Schrödinger, and others explored them. At some point after sufficient "explication" has taken place, scientists will inevitably discover that beneath their new explicate layer lie yet new mysterious implicate layers and the previously explicate (or some of it, at least) will now become implicate, like the dye drop winding back into the fluid. Movements that at one layer appear to be random at another will be found to have a complex, unfolding order. The universe will never be explained, but scientists will have the pleasure of winding round and about never-endingly into it.

But we'd best come up a bit to lighter waters. Back at the explicate level where most contemporary physicists are still struggling, we have the nagging phenomena of quantum mechanics. What can Bohm's implicate model tell us about these?

Bohm's theory permits the electron and other particles to be viewed from an entirely new perspective. In the current mechanistic physics, the electron is thought of as a separate particle which exists at each moment in only a small region of space and which changes its position with time. In Bohm's model, the electron is a total set of ensembles enfolded throughout the whole, not localized in any particular spot in space. At any moment one of these ensembles may *un*fold and make a click or a track at a detector, but the next moment the ensemble may *en*fold (wind back), to be replaced by the ensemble that follows, like the successive dye drops that had been wound in at the same place in the glycerine cylinder. The single particle appearing to our senses through our instruments is an abstraction of an underlying process of undivided movement. Bohm's picture is bizarre. In one sense the particle is not one thing at all; it is successive unfoldments. But in another sense, since everything is enfolded into everything else, the particle is always the same!

Bohm's implicate order neatly accounts for a universe that appears both continuous and discontinuous. It just depends on how the ensembles unfold. If they unfold one after the other very near each other, they look like a single particle moving continuously across space. If they unfold with greater apparent separation, they appear like a particle jumping discontinuously from one place to another or even like a particle separating into several other particles and then reemerging as itself again.

Bohm's hypothesis also accounts for the wave-particle duality that so troubled early quantum physicists. If obstacles are placed in the way of an ensemble as it unfolds, the ensemble will manifest a different explicit order. This would be like placing a fine wire in the way of the unfolding dye drop. In the experimental situation, the wire is the scientist's observing apparatus. By changing apparatus, the scientist can determine whether electron ensembles appear as particles or waves.

Now for an additional mind-boggling wrinkle: In Bohm's implicate universe *both the observing apparatus and the observer himself are also unfolding ensembles.* The click on a detector, the interference pattern on a photographic plate, or the track in a bubble chamber must therefore be seen as a kind of flowing intersection between (A) the unfolding implicate ensemble that constitutes the observer and his apparatus (and the theories which lie behind both) and (B) the unfolding ensemble called the particle or wave. A and B are like two vortexes. "Facts" or "observations" appear, as it were, at the "edge" of these vortexes where they merge.

MODEL C: A HIGHER-DIMENSIONAL FISHTANK

Bohm's third major analogy or model broadens the conception of his implicate order still further. It also conveys principle number three of looking-glass wholeness. It goes like this: The universe doesn't exist in only our familiar three dimensions or Einstein's four. *It is a universe of countless dimensions which embody its wholeness.*

In the quantum context, this third analogy provides a good way to understand the strange effect of nonlocal causation raised by the EPR thought experiment and the later actual experiments on the Bell hypothesis (page 88).

Imagine a fishtank with two television cameras set at right angles to each other, their lenses pointed to the tank. The images from the cameras are projected into two television sets A and B. A fish swims by and faces into the lens of camera A. Television A shows the fish head on. Television B shows a side view. Now imagine you didn't know about the TV cameras or the fishtank and consider the fish as elementary particles. What would you think about the relationship between the pictures on the two screens? Probably you would conclude that these were two particlefish interacting in some way. You would notice, for example, that when particlefish A turns through a right angle, particlefish B turns the other way. Perhaps this is some correlation of particlefish spin?

The rabbit thinks the two-dimensional fish A and B are separate things somehow correlated. He doesn't realize that they're actually projections of a three-dimensional world in which A and B are one.

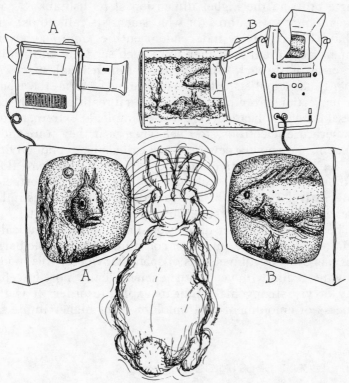

If a scientist is clever he might come up with a theory which fairly accurately describes and predicts a relationship between the two particlefish, though his theory would be based on an illusion. There aren't two fish causally related at all. The difficulty is that he sees the fish in only two dimensions (the two-dimensional TV screens), though the reality from which these projections come is a "higher-dimensional reality." It is a reality with three dimensions. In that three-dimensional reality there is only one fish, a single actuality. This three-dimensional ground holds the two-dimensional, apparently correlated projections within it, but is an essentially *different* reality.

Bohm likens this situation to the EPR experiment. The difference is the number of dimensions involved. In the EPR, each of the twin particles has three spatial dimensions. Together they have six. When the twins are quite distant from each other, they still appear—in our three-dimensional reality—to be correlated, like the particlefish. We see the correlation but we can't find the causal connection. How *does* one affect the other across such distance? The answer, Bohm says, lies in a higher dimension—the sixth dimension, to be precise. In that higher-dimensional reality we would realize that two separate things are a single unity, as the particlefish is a single unity in the higher dimension of its fishtank. Sometimes, Bohm says, atoms (or what scientists call atoms) will behave like relatively independent entities and it's convenient to treat them as if they were such separate entities interacting with each other in three-dimensional space. At other times scientists will have to face the fact that the atoms are projections from higher-dimensional realities, that is, expressions of implicate orders. For example, at extremely low temperature, electrons stop behaving as if they were independent and show a new property called superconductivity. In the superconductive state, electrical current can flow without resistance. The electrons will go around obstacles cooperatively without being scattered or diffused. A similar phenomenon occurs in superfluidity and in laser light.

The number of higher-dimensional realities is theoretically infinite. For example, an object containing 10^{24} atomic particles has 3×10^{24} dimensions of space. Obviously this model provides Bohm with yet another, scientifically more forceful, way of visualizing his implicate-explicate order: It is the process of enfoldment and unfoldment of higher implicate dimensions into our own familiar explicate order of three-dimensional space.

The hologram, which is constructed by an electromagnetic field, obeys the laws of quantum mechanics, and the mathematics of quantum mechanics relies on the concept of a multidimensional reality. Bohm shows how this mathematical system, which physicists generally treat only as an abstract formulation, gives insight into something that is "real." He is thus able to supply a kind of picture of subatomic reality which quantum mechanics lacks, though Bohm would be quick to point out that even this idea of a multidimensional reality would ultimately have to be treated as an abstraction—a higher-dimensional abstraction.

For Bohm, the multidimensional reality is one unbroken whole extending through the universe and including all of what scientists call particles and fields. The holomovement enfolds and unfolds in a multidimensional order. In this multidimensional order, subtotals can sometimes be abstracted and studied.

Bohm also treats time as a projection from a higher-dimensional reality. Classical mechanics and relativity consider time a primary dimension. In relativity, one feature of time, the speed of light, is taken as a constant. In classical mechanics, time is an absoute dimension, one of the axes on the Cartesian grid. A Bohm innovation is to treat time as secondary. For him, both time and space are projections from a higher-dimensional reality. Another way to put this: Space and time are subtotals which unfold the way the dye-drop "particles" unfold.

All of us have had direct experience of what Bohm means. Sometimes you meet someone you have not seen for a long time (as measured by the clock), yet it seems as if no time has passed. It is as if that time you were in with your friend went at a different pace, unfolded differently, from the time, say, that you spend at the office. Some activities are so absorbing that we may feel no time or very little has passed and then we look up at the clock to discover we have been at this thing for several hours. If we are bored, time seems to unfold with excruciating slowness. Bohm says that these different time orders are as "real" as those ticked off mechanically by our wristwatch or the schoolroom clock.

Time orders for physical systems unfold at different rates too. In the decay of a radioactive atom, a previously inert nucleus will suddenly emit a particle after a second, an hour, several years. A scientist may predict a 50:50 chance that the atom will decay in a given clock time, yet the individual event is discrete and totally unpredictable. Clearly a linear, clock time is inappropriate for the description of this event which unwinds "in its own time."

PHYSICS AND BEYOND

Bohm's description of the "reality" of psychological time orders and his concern with the immediate experience of movement show that the implicate-order theory of wholeness, by its very nature, extends outside of what is usually considered the domain of physics. For example:

The theories of mechanistic biology currently espoused by most scientists assert that life appeared as the result of a random encounter of nonliving molecules. Life was sparked into being, according to one hypothesis, by an early electrical storm. In Bohm's holistic universe there are no such random events. Life, he says, is implicit in what we call inanimate matter. For the most immediately obvious sense of this, consider a flower or weed. The weed is an explicate manifestation of the ensemble of all the atoms in the environment which drew together (like the threads of the dye drop) to form the seed, the air, and the mineral nourishment that made the weed spring up. At some unfolding point all these things gathered into the weed.

In a larger and perhaps less obvious sense, Bohm says the holomovement itself must include the principle of life. Life was no accident. Life as a whole is a subtotal, as is nonliving, inanimate matter. But these two subtotals are only relatively separate. In fact, life and nonlife constantly intermingle and enrich each other. When the weed dies it will rot and increase the store of inanimate matter, which, in turn, will unfold, giving birth to the next form of life. We will get a particularly strong picture of this intermingling of subtotals in the next section with the "dissipative structure" theory of Ilya Prigogine.

Bohm believes distinctions between the animate and inanimate are only abstractions, useful in some contexts but ultimately inaccurate. If he had to choose, he would say that the statement "Everything in the universe is alive" is a more useful description of reality than the one which currently prevails in science. The current mechanistic theories see the universe as fundamentally controlled by blind, inert matter and the chance collision of particles. For Bohm, even the chemicals of the soil are living.

Bohm applies essentially the same argument to consciousness as he does to life: Consciousness also is implicit in the holomovement and therefore implicit in all matter, and in that sense is "contained" in all matter. In Part 5 we'll look at the detailed proposal and experimental evidence of neurophysiologist Karl Pribram that consciousness takes place in the brain holographically. Pribram's theory makes the relationship of consciousness to the universe something like that of a hologram looking at a hologram—the two together making a larger hologram. But we should first look at Bohm's view of consciousness from his unique perspective of the implicate universe at large.

As Einstein brought us the space-time continuum, seeing space and time as one inextricably linked process, Bohm brings us the matter-mind continuum, seeing consciousness and matter as inextricably linked. The issue Bohm takes up here is an old one. As we've said, Descartes described it most distinctly when he noted that matter and consciousness seemed to be two very different orders. He called matter "extended substance" to indicate that it is "things" existing separately in space, what Bohm calls the explicate order. Descartes called consciousness the "thinking substance." Consciousness is not really extended or separated in space, making it closer to what Bohm is calling the implicate order. How can these two very different orders have any relationship? It seems as unlikely as marriage between a horse and a butterfly.

Descartes solved the problem with God, who stood outside of both matter and consciousness and created them—the marriage was made in heaven. Science dropped the idea of God, so the problem reemerged. How can these two very different orders get along?

Current notions of this relationship are self-contradictory. On the one hand, scientists believe the order of consciousness is essentially separate from the order of matter. This is evident in the very concept of a scientific experiment where it is assumed that the observer can stand apart from the observed. The discovery of quantum wholeness has not substantially altered most scientists' belief in objectivity, at least on the macro (cat) level. On the other hand, scientists also believe that consciousness is nothing more than matter—an electrochemical phenomenon. Bohm's theory resolves this paradox ingeniously.

First, as the holographic model shows, light, electromagnetism, sound, all the energies, enfold information about the whole universe in each region of space. Thus, as these energies enter consciousness through the sense organs, it is, in each instant, the whole which consciousness encounters, which consciousness (and perception) *is*. Moreover, as Pribram's hypothesis will claim, these energies become translated by the sense apparatus and recorded through the brain holographically.

Second, since the brain and sensory apparatus are themselves composed of matter, which is also waves, the very material and processes of the brain are a holographic imprint of the whole. Therefore, both the order of consciousness and the order of matter, observer and observed, are projections and expressions of the implicate order where the two are one and the same. Each is a mirror reflecting itself. Mind is a subtle form of matter, matter is a grosser form of mind.

The next question is: Why is it that the explicate order of apparently separate things is what we experience in consciousness? The immediate answer is that it's not always what we experience. Children, members of "primitive" cultures, artists, mystics, and virtually all others at some time or another experience the implicate order. In the case of artists they find ways to give the implicate, which they usually call "intuition" or inspiration, an explicate shape. In the case of children, they grow up. So the question is narrowed to: Why is it that sometimes or most of the time we experience the explicate order as the only reality?

The explicate is composed of subtotals, relatively stable forms that are expressions of the underlying implicate order. Here we need to remain acutely aware that the implicate and explicate are not really different. They're two forms of the same thing. What we call the explicate (things) are simply more relatively stable than what are not things: The Gulf Stream is merely a more relatively stable form in the general movement of waters in the North Atlantic. Now consider another subtotal, a track left by a dinosaur. Erosion will transform it, but very slowly. It remains relatively stable for a very long time. The track is a kind of memory in matter. Memories in consciousness are like dinosaur footprints, relatively stable forms in matter, though certainly not quite as stable as dinosaur footprints. Brain cells and brain energies are immeasurably more plastic than stone, transform more quickly. A person may remember her third birthday party differently when she's sixteen than when she's sixty, but, by and large, relatively stable forms will appear. In fact, as Piaget has shown, memory is trained in the childhood years to form these stable images. To give a simple example, a child learns that the word "chair" covers a wide variety of forms which have some relatively invariant elements. This is the process of abstraction described before as mapmaking. Eventually the child learns to "see" a chair as a chair and recognize it in many different contexts. (Remember Kuhn's example of Johnny learning to see swans?) This "seeing" is sensory information screened through the relatively stable forms in memory. So there is a tendency to "see" the relatively stable forms (subtotals) in the sensory input which fit with the subtotals in memory.

Of course, this process is only "relatively" stable and so is actually constantly transforming and fluid, which is how we learn anything new.

The stable forms recorded in memory as the screens or guides for seeing differ from place to place and culture to culture and from one era to another. One culture builds up a memory screen which abstracts out of the flow of nature stabilities such as the different wavelengths of energy, another culture has a screen which abstracts thirty-three kinds of snow.

So ordinary consciousness responds to the explicate because that consciousness has been trained through acculturation to consider itself an explicate order, to screen out and suppress vast dimensions of its own implicate nature. One of the explicate-order forms which consciousness adopts is the sense of personal identity or self. Humans come to think of the individual self as a fundamentally separate "thing" which persists despite an immensity of changes that take place in one's life.

Becoming an explicate order mirroring the explicate orders in the universe has great advantages for consciousness. This faculty of mind literally allows us to go to the moon. But it also has meant that we learn to ignore the transitory, subtle features of existence, the nuances and differences of things that memory has not been trained to regard as stable, including the nuances and subtleties of our identities. As a result, the relatively stable aspects of the unbroken whole come to be seen as actually separate parts, deadly and final realities. Among these lie our selves.

Bohm believes that there is a grave fallacy lurking the concept of individual consciousness. In the implicate order, consciousness as a whole—the total consciousness of humankind—has a more primary reality. Even more deeply than this, all of consciousness is enfolded in matter and matter is the unfolding of consciousness. Thus individual consciousness, like an individual electron, is an abstraction. A useful one at times; at others, destructive and confusing.

Bohm's implicate-order approach resolves some sticky scientific problems with the concept of consciousness.

One is the brain-mind or mind-body problem. Scientists have argued for many years over whether the mind is limited to the brain. The mechanists have insisted that it is. Proof: When the brain dies, the mind dies. Others, called "vitalists," have argued that the brain is an expression of the mind which extends beyond it. Proof: Parts of the brain can be destroyed but the mind remains intact. Again, Pribram will have an illuminating perspective on these matters. But Bohm's approach is a prelude and overview; it shows brain and mind, mind and body, enfolding each other. They are neither separate nor the same. Like two images of the single fish, they are projections of a higher-dimensional reality. Mechanists like famed behavioral psychologist B. F. Skinner have argued that the mind is nothing more than a stimulus-response device with a complex wiring diagram. Bohm responds: "Physics has shown that the mechanistic order doesn't fit experience, and if it were going to work any place at all it should be in physics. Still less does it work in the field of mind. Actually in this field it works mainly in certain rather limited areas such as teaching pigeons to peck in a certain sequence."[3]

Another problem resolved here is a familiar one, the nemesis of quantum physics: What is the relation of the observer to the observed?

In classical physics the observer was separate from the observed. They were separate parts of the universe. With quantum mechanics, Heisenberg first barred the observer from certain aspects of what he observed and then the Schrödinger cat problem excited some theorists to propose that the observer actually affects what is observed by collapsing the wave function. Both these views subtly retain the classical idea that the observer is separate, though in the second case he and what he observes are "interacting." For Bohm such ideas completely distort and inflate the role of the observer and lead to confusion because they are fragmentary. For him, both observer and observed appear from the same underlying indivisible process and flow into and out of each other like the stream through vortexes. The division between the observer and observed is a sometimes convenient abstraction which permits some deeper observation to take place. Notice the word "observation" in this sentence doesn't imply who or what is doing the observing. The observer is not causing the observed. They are both in a sense causing each other and being caused by the underlying whole movement. By saying it this way we can see Bohm has changed the whole idea of causality from a chain of events to a complex picture where effects and causes interweave with each other. Schrödinger's cat dilemma dissolves in the completely new light where decaying atoms, observers, wave functions, and cat are intermingling orders all present in different degrees of enfoldment.

In this joining of consciousness with matter, Bohm's theory reveals itself as admirably self-consistent. When he talks about consciousness having "insight" into the implicate order or any other order—Heisenberg's insight on Helgoland into the world of the quantum, Einstein's insight into relativity—the word "insight" is synonymous with a leap to the implicate level. The insight may quickly take an explicate form (a poem, a theory, a sigh), but Bohm wants to show us that explicate expressions (be they scientific theories, poems, or sighs) do not eliminate the implicate. It is still there, behind everything, slipping steadily away from attempts to explicate it fixedly, like sand pulled out from beneath one's feet by an undertow. Where does insight come from? Bohm would say from the holomovement and perhaps something even beyond the holomovement (which is, after all, only an explicate idea of an implicate process). An insight is not Bohm's insight or Heisenberg's or Leonardo DaVinci's. It is the movement of the whole expressing itself through explicate forms.

A COSMOLOGICAL N-DIMENSIONAL MOVIE

We will close this brief sightseeing tour of Bohm's universe with a series of zoom shots—shots showing what physicists call cosmology. As the name suggests, "cosmology" means the big picture. And this one is even bigger than big because it's the cosmology of the implicate order.

When physicists applied quantum theory to empty space, they discovered that there is a minimum amount of energy in any region. Calculations of this minimum energy show two amazing facts: (1) Space and therefore time at this minimum-energy level become totally undefinable. (2) *In one cubic centimeter of empty space the amount of energy is much greater than the total amount of energy of all the matter in the known universe!*

Empty space therefore is not empty at all; it's full, an immense sea of energy on top of which matter as we know it is only a "small quantized wavelike excitation . . . rather like a tiny ripple."[7]

It should be said that some physicists believe this infinite sea of energy is an illusion, an error in the mathematical foundations of quantum theory. Bohm, however, takes it seriously. He believes that in general contemporary physicists ignore this immense background energy because they're interested in matter. It is rather like focusing so much on a crack in a blank wall that you forget there is a wall; the vast sea of energy is subtracted out of physicists' equations.

But Bohm believes the relationship between matter and this sea is crucial. He suggests it by an analogy: At absolute zero a crystal will allow electrons to pass through it without scattering them. They go through as if the space were empty. But if the temperature is raised, flaws appear in the crystal and scatter the electrons. From the electrons' point of view, what appears as "matter" are the flaws in the crystal; the rest seems empty space. It would appear that these flaws exist independently and that the main body of the crystal is sheer nothingness. The universe we inhabit is like the crystal. Its "flaws" are the matter of galaxies, planets, our selves. The nothingness and matter are as inseparable as the vortex and the stream. This crystal sea of energy nothingness is the multidimensional implicate order. Thus:

The entire universe of matter as we generally observe it is to be treated as a comparatively small pattern of excitation [on the energy sea]. This excitation pattern is relatively autonomous and gives rise to approximately recurrent, stable and separable projections into a three-dimensional explicate order of manifestation, which is more or less equivalent to that of space as we commonly experience it.[7]

Bohm therefore regards the so-called "big-bang" explosion which is supposed to have given rise to our universe as more like a "little ripple" on the energy sea. He compares it to what happens in the middle of an ocean where a myriad of small waves may come together with their phase relationships arranged so they produce a very high wave that seems to appear out of nowhere. As this wave explodes outward it breaks up with smaller ripples and constitutes an expanding universe. This exploding universe would have its own space folded in it.*

Bohm thinks the effort by contemporary physicists to treat the universe as if it existed independently of the implicate sea of energy on which it formed will eventually lead to confusions. Already the notion of black holes suggests a connection to this deeper cosmic-energy background.

But the cosmic background energy is itself not the end. The calculations suggest that it is possible that beyond this sea may lie a further domain or set of domains at present unimaginable. These could be extensions of the implicate order or they could involve yet undreamed notions of order.

So the camera begins with a small, frail flutter, barely a scintillation, on some fluid surface. This minuscule faint flutter of movement is consciousness. The camera pulls back and we see that the shimmer of consciousness was folded in a slightly larger but still quite small movement—matter. But where is matter? As the camera zooms away we see it is a ripple in a vast ocean of energy itself on a vaster as yet darkly perceived movement which may extend into yet further and unthought-of dimensions.

But now we see that our imaginary camera's journey has not only been immense, it has been more than strange, for in

* Physicists have determined that though the universe is expanding there is no actual center from which this expansion is taking place, or, more accurately, the center is everywhere. At any point in the universe, it is expanding in all directions.

every sweep of its motion it gazed into the looking-glass. Wherever it gazed, in whatever it gazed at, all else was reflected.

7. Languages of Wholeness

Throughout his writings and talks Neils Bohr pointed to the deep and subtle relationships that exist between the formal and informal languages of science. A physicist, for example, may spend the day manipulating abstract mathematical formulae but during his coffee break will communicate insights, conclusions, and hunches to his colleagues in everyday language. Science is a social activity that involves both individual research as well as communication among the whole scientific community. The development of new ideas and scientific concepts therefore involves complex interactions on the levels of mathematics, logic, and ordinary language.

Most scientists take extreme care with their mathematics yet few would be as deeply concerned with the language they employ during blackboard discussions or scientific lectures. Bohr emphasized again and again the confusions and paradoxes that occur when everyday language is pressed into the task of constructing concepts in the quantum world. Our language has evolved for dealing with the macroscopic world of relatively autonomous objects—"things" like stones, chairs and houses—and is inappropriate when applied to electrons, transitions, and quantum processes.

David Bohm is similarly concerned with the relationships that exist between the formal and informal languages of science and the difficulty of communicating scientific concepts. Bohm has, for example, studied the conversations and exchanges between Bohr and Einstein and concluded that the two men used words like "interaction," "object," and "signal" in very different ways and so failed to come together in their dialogues. Mathematical expression and ordinary expression, Bohm believes, are linked and inseparable. For Bohm, all levels of discourse about science, or anything, must be in harmony or confusion will result. The important thing, Bohm thinks, is to pay very close attention to the whole process of theory-making: the nature of the theory itself, the nature of the experiments, the nature of the mathematical form, and the nature of informal language used to express it.

THE RHEOMODE

In the realm of "informal" discourse, Bohm has gone so far as to invent his own language to express the implicate order. He calls it the "rheomode" from the Greek *rheos* meaning "flowing." With the rheomode Bohm attempts to overcome the subject-verb-object fragmentation of most languages. Take a simple example of this fragmentation: A cat and mouse rush past you in a flurry of activity. We would say, "The cat chases the mouse." Enfolded within this simple sentence is an entire world view. It begins with the nouns "cat" and "mouse"—separate objects in the universe—passive, autonomous existences. The verb "chases" is an action separate from these objects, implying, among other things, that the action is done on the mouse by the cat. Yet the whole action is more complex. It is a dance of death and life in which the cat and mouse are inescapably embedded. Bohm attempts to overcome these artificial separations by making all words in his language variations of the verb. For instance, he uses the verb "to levate," meaning to lift something into attention in such a way as to include the question of whether what you're talking about is relevant to the context, at the same time calling attention to the fact that you're calling attention to something—that to some extent you're separating it out of the whole by the tweezers of language. The verb "re-levate" means to do this all again, including the whole process of considering whether what you're saying is relevant "this time." Other verbs and variations similarly call attention to the process of language, thereby, Bohm hopes, avoiding the fragmentation language often brings and, instead, simulating in language a sense of flowing movement. Bohm has even gotten some of his graduate students to talk in rheomode, albeit with tongue in cheek. But if its reception has been less than enthusiastic, the rheomode at least demonstrates Bohm's commitment to trying for harmony at all levels.

SOME ANTI-CARTESIAN MAPS, PLEASE

Considerably more effort and time has been put into developing new mathematical approaches to the holomovement. Here the stakes are higher. If Bohm's theory is to succeed in the scientific community, it will have to attract the puzzle solvers of normal science with a formalism that allows them to perform experiments and investigate the implicate order in detail. In other words, the implicate-order theory will have to speak eloquently in the mathematical language of science or it will have to persuade scientists to speak its language.

Up to this point we have discussed Bohm's metaphysics, his world view of twentieth-century science. But as this view developed and its expression crystallized, Bohm was also engaged in active research. His publications may not have been as numerous as those of other scientists discussed in this book, but so many of his ideas have been significant and influential that no matter how "far out" they seemed, as a physicist Bohm earned the respect of his peers.

Over twenty years ago, Bohm realized that the Cartesian order had remained an essential part of science during the past two hundred years. According to Descartes, points in space can be represented by numbers. Lines or paths in space are represented by infinite sets of points or by algebraic equations. This insight of an essential unity between algebra and geometry was a powerful and valuable one. Geometry, which describes the relationships between objects and shapes in space, was equivalent or could be represented by the algebraic ordering of numbers and the relationships between functions.

This Cartesian order enabled Newton and Leibniz to invent the calculus, and today it seems impossible to imagine a science that does not rely on integrals, differential equations, and functions of space-time variables. Some physicists, however, questioned the use of continuous space-time variables in what was otherwise a theory of discontinuities and quantum jumps. In the early years of quantum theory a few physicists speculated about things like quantized time or space which took the form of a discontinuous lattice. More recently, as mentioned, John Wheeler has suggested that at very small distances time-space breaks up into a foamlike structure and that the everyday properties of space-time must be derived from something more primitive—a "pregeometry."

Bohm is particularly emphatic that Cartesian description is incompatible with the insights of quantum theory. A new description is needed, he thinks, which reflects the holomovement and implicate order, yet can reduce to the space-time of relativistic physics at an appropriate level.

While at Bristol University in the late 1950s, Bohm was investigating various topological ideas to see if they could be applied to quantum systems. Topology is a logically more primitive form of geometry which makes use of such relations as "inside," "outside," "enclosed by," "intersected by," and so on. Topology could be thought of as a geometry performed on a sheet of rubber that is constantly being stretched from different directions. As the sheet is distorted, a square becomes a circle, then an ellipse, next a triangle. Familiar geometric forms flow into each other and deform, yet, provided the sheet does not rip and tear, certain relationships are preserved. Two circles which intersect will continue to intersect even if they are transformed into squares. No matter how hard one stretches it, a line remains a line and cannot become a closed figure such as a triangle. If one figure encloses another then it will continue to do so no matter how much the underlying space is distorted.

Mathematically these ideas can of course be extended from flat figures on a two-dimensional sheet to more general forms on an abstract space of any number of dimensions. Bohm suggested that these topological relationships may be more natural as a mathematical basis for quantum physics than the Cartesian order of space-time. Indeed, topological order is closer to the way we interpret the space and time around us than are continuous coordinates. Piaget has shown, for example, that young children can distinguish intersecting figures from those that do not intersect long before they can distinguish triangles from squares or circles.

In the 1960s, after Bohm moved to the University of London's Birkbeck College and began developing the idea of holomovement, he was aided in the formal development of this research by Basil Hiley, a physicist who has been his colleague for the last twenty years. The collaboration has proved a highly effective one. In a personal letter to the authors Hiley wrote, "My main driving force is that I want to understand physical process and how best to express our insights in some form of mathematical structure." Hiley is a quiet man who follows his own paths, yet both researchers share a common perception at the deepest level of their work in physics.

Hiley's strength lies in translating physical ideas and insights into mathematical form, and he believes that mathematics is the universal language of communication among physicists. Bohm, by contrast, is often impatient with the business of mathematics and proofs. He has been heard to say, "Why do people spend so much time proving what is obvious?" Bohm can often see to the heart of an idea and the shape of the final mathematical expressions needed, but balks at going through the details of a mathematical argument and checking all the steps to obtain the most elegant approach. He is one of those few physicists who have the ability to throw down a rapid draft of a proof or mathematical derivation guided by physical insight. On closer examination the mathematics may contain minor slips and errors but the final result will be perfectly correct. Scientists of such ability can be quite disconcerting at a seminar. The visiting lecturer will begin his talk and may set down some equations which he intends to develop over the next hour. A scientist like Bohm may react by saying, "Well, that isn't going to work," because he immediately sees the implications of the arguments that are to follow.

Throughout the 1960s, Bohm and Hiley explored the mathematical possibilities of using topology in quantum theory. By the end of the decade they had settled on co-homology as the best approach. This is a form of topology that does away with an underlying space-time structure. It is as if the rubber sheet on which various geometric figures were drawn were discarded but the abstract mathematical relations were retained. Bohm and Hiley argued that with these relationships alone it should be possible to express the fundamental laws of physics.

In the nineteenth century, Michael Faraday had attempted to express his findings on the phenomena of electricity and magnetism in what today looks like a mathematics very close to co-homology. Faraday talked of "lines of force"—lines of magnetism that run from one magnetic pole to another. Every line that leaves one pole must complete the circuit and enter the opposing pole. The stronger the magnet, the more lines that complete the circuit. An electrical current is another form of circuit, for the current leaves one pole of the battery and makes its way through the electrical connections and back to the opposite pole. Accompanying this electrical current will be a magnetic field represented by other lines of force that intersect the circuit. Hence the phenomena of electricity and magnetism are represented in terms of circuits or closed lines, some of which intersect others. Physical quantities are represented in this scheme in terms of counting the number of lines or observing if some lines are enclosed by the circuits made by others.

Although Faraday didn't realize it, these ideas are essentially topological relationships and indicate that the laws of electromagnetism can be expressed in co-homological mathematics. The formal approach is totally different from that of Maxwell, who treated electromagnetism in terms of fields defined in a continuous space and time. Maxwell's theory makes use of differential equations and continuous quantities that are defined at each space-time point.

Bohm and Hiley showed that the major areas of physics—electromagnetism, thermodynamics, quantum theory, and so on—could be expressed in co-homological terms, that is, without any need for an underlying space-time at all but simply using the logical relations of topology. In a sense this approach goes back to Leibniz, the contemporary of Newton, who criticized the notions of absolute space and time and suggested that these orders were instead simply the relationships that exist between material bodies.

This work from Birkbeck College was not, however, a "new theory of physics" but simply showed that existing theories could be cast in a totally different form. The underlying motivation was his realization that Cartesian order is incompatible with the expression of quantum theory and that by casting the laws of physics in a topological mathematics, their deeper structure could be more clearly revealed. The expression in terms of co-homology is an intermediate step, for it deals with physics that does not refer to a continuous, underlying space-time. Perhaps at some still more primitive level of "pregeometry" these topological notions will themselves turn out to be derivative.

During the last decade, Bohm has concentrated on developing and communicating his ideas about science, consciousness, and the implicate order. His formal publications have included a further extension of his early hidden-variable proposal, this time including some detailed suggestions for experiments to actually detect subquantum (implicate) effects in quantum situations involving very high energies and short distances.

With some of his students and co-workers he has also engaged in a reexamination of an idea from the early days of quantum theory. At a conference in 1927, Louis de Broglie suggested that individual, localized electrons existed and that they were "guided" or directed by some overall "potential." It was this potential "guide wave," *not* probability waves, that was the solution to Schrödinger's equation.

De Broglie's theory was severely criticized, and he dropped it until, in 1953, Bohm came up with essentially the same idea. Bohm and de Broglie later published together on the "pilot-wave" theory, but real progress was not made until the early 1980s.

The guide wave is viewed as an expression of the total experimental arrangement. Each part of the apparatus contributes in some complex way to the overall pattern of the wave. The wave guides each electron the way a radar signal guides an airplane.

Bohm insists this approach is only a provisional point of view which offers new insights into the mathematics first proposed by Schrödinger. On the one hand, the electron appears as a definite particle, a viewpoint which might have offended Bohr. On the other, the "wholeness" of the experimental situation, which Bohr constantly stressed, now emerges as a necessary consequence of the mathematics.

The pilot-wave theory has been applied to the double-slit experiment. Here the quantum potential causes particle tracks to bunch together as they pass through the slits, accounting for the familiar interference fringes. The guide wave acts nonlocally to organize the arrangement of every particle. In this way it is possible for a system to evolve into separate "parts." Quantum measurements, from this point of view, are quite objective and have no need for a human observer and his consciousness. The Schrödinger's cat experiment has a definite outcome before the box lid is opened because the whole quantum system, including the observer, has already been organized by the guide-wave potential.

ALGEBRAS OF THOUGHT

Bohm has also carried out a sustained search for a formal, mathematical expression of the holomovement. The ideas of co-homology and topology provided certain insights but, in retrospect, did not go far enough. Within the last two or three years a new approach has begun, based on the mathematics of the nineteenth-century mathematician H. G. Grassmann.

Basil Hiley had been invited to give a lecture at a series of seminars titled "Space and Time." During the discussion that followed the lecture an American graduate student asked him if he had read Grassmann's book, known in English as *The Theory of Linear Extension, a New Branch of Mathematics*. Hiley began to study it and realized that it had an important bearing on expressing the holomovement.

Herman Gunther Grassmann was born in Stettin, in what is now Poland, in 1809. He was a mathematician who also took a deep interest in linguistics, botany, physics, and folklore. His major mathematical work was an attempt to develop some ideas of Leibniz's involving an algebra of geometrical relations. Ths work was an effort to escape from the straightjacket of three-dimensional geometry and Cartesian order. His contemporary in Dublin, the eccentric W. R. Hamilton, worked along similar lines.

Mathematics, Grassmann and Hamilton believed, is not so much about orders in the material world but about the processes of thought. Mathematical objects are the abstract constructions of the human mind; their relationships and ordering are the true business of mathematicians. If a particular branch of mathematics later has application in physics or engineering, then this is purely incidental—what Eugene Wigner has called "the unreasonable effectiveness of mathematics." Bohm, of course, would consider such a mirroring of conciousness and matter an expression of the holomovement. An effect of the looking-glass.

Grassmann directed his attention to the movement of thought and the way one thought flows from another. To the mathematician this was unlike the ordered relationship of points on a line or the successive points in space occupied by a moving object. Thoughts are inseparable; one flows from the other, each contains the implication of the other. Thoughts are like the poles of a moving relationship. Pursuing this insight, Grassmann began to develop "an algebra of becoming"—a mathematical expression which has much in common with Bohm's holomovement.

Grassmann's new algebra was not well received by his contemporaries, so, in his mid-fifties, he turned to the study of Sanskrit and published a dictionary to the *Rig Veda*. It remained for the English mathematician William Kingdon Clifford, born in Exeter in 1845, to extend the ideas of Grassmann and Hamilton. Today the work of Clifford, Grassmann, and Hamilton is considered mathematically important, but only certain aspects of their work have come into fashion. In a sense the "static" aspects are considered important while the treatment of "becoming" has been forgotten. To Basil Hiley, however, it is Grassmann's original insight that is most intriguing and valuable.

In the first years of the 1980s, Bohm and Hiley are exploring the applications of the Clifford and Grassmann algebras in search of a pregeometry which expresses the insights of the implicate order and reduces, in appropriate limits, to the established results of quantum theory. This research has only just begun. By cutting away the reliance on Cartesian order and an underlying space-time structure, it may be possible to derive a theory which contains both relativity and quantum theory as subsets. The final expression of such a theory, however, lies in the future.

Two other avenues of mathematical research have also been explored. One involves current interest in what are called "solitons." Solitons are mathematical treatments of things like vortexes and waves which allow them to be described semi-independently—abstracted, as it were, mathematically from the flow.

FRAMING REALITY

The other avenue involves a revival in a new form of Lorentz's ideas of material frames (see page 57). For Lorentz, material frames—such as the frame of the planet earth—are held together by electromagnetic forces. They define distance and time. In Lorentz's case, these frames were embedded in an absolute space and time. Bohm visualizes them in relative terms and believes they may account for the explicate structure of space-time itself.

To illustrate how a frame works, let's consider a collection of clocks. In a jeweler's shop the clocks and watches all tick away on their own, showing slightly different times. But suppose that these are all electric clocks and they are connected together. Such clocks will couple together; they will influence each other. If one runs slightly faster, the others will slow it down until all the clocks tick at exactly the same rate and show the same time. Mechanical systems can exhibit similar phenomena; when they are unconnected their oscillations may fluctuate, but if they are coupled together they begin to beat in unison and exhibit a common rate.

Bohm's idea is that at a certain level subatomic systems will couple together to produce a common order.*

Suppose that elementary particles all possess their own primitive pregeometry—in a sense, their own primitive space-time. At this level, space and time become a "multiplex" of enormously high dimension. But if the various particles couple together they may produce a common order that we recognize as our familiar space and time.

In classical mechanics the notion of material bodies can also be represented in an abstract way in terms of waves— technically, Hamilton-Jacobi waves. Conventionally these waves are not thought of as having a real existence but are simply used as a device for calculation. Bohm, however, thinks this approach suggests a reality in which waves couple together to produce beats and these beats define a time ordering for the system.

In conventional relativity, observers who move at different velocities perceive different space-time orderings. Bohm thinks that different space-time orderings might be produced as the matter-waves couple together in different ways. Material frames which have different orderings are then interpreted as having different motions with respect to each other. At present these ideas are hints to some larger picture. They suggest that the ideas of inertia, straight-line motion, velocity, acceleration, and space-time can all be derived from some basic ordering of coupled matter-waves.

Bohm is a man capable of giving off a host of ideas. In the words of one physicist, "Some are brilliant, many are obscure, and some are just plain nonsense." The idea of material frames may well prove important in the future, for it appears to contain a germ of a valuable insight into what Bohm calls the laws of absolute necessity, which bind together different unfolding ensembles in the implicate order.

---

* It should be said that Bohm makes this proposal from a conventional point of view in which systems are considered as fragments. The illustration does not begin from the ground of holomovement at this point.

On the other hand, when its full implications are worked out mathematically, the idea may prove incorrect or of only limited usefulness.

In attempting to develop these languages of wholeness, Bohm and his colleagues have made a rich and varied start but obviously have a long, perhaps impossibly long, way to go.

8. Looking Back from the Other Side

About 500 B.C., the pre-Socratic philosopher Heraclitus said: "Everything flows." However, aside from some cryptic and thought-watering comments about the unity of contraries, only a few fragments of his explanation (if he made one) of exactly *how* things flow have survived. David Bohm has begun the arduous and delicate task of giving scientific form and detail to this ancient statement.

It seems a poignant irony that Einstein had once remarked he thought Bohm would one day be the person to solve the unified field-theory problem which eluded Einstein himself in his last years. It could be said that Bohm has indeed solved that problem, though in a way that might well have dismayed his classically trained mentor.

At the beginnings of the quantum revolution, Niels Bohr asked: Why is matter stable? To answer this question he was forced to conclude that electrons don't fall into the atomic nucleus because they can only occupy certain discrete orbits. After that things became more complicated, and Bohr's question continued to vibrate behind the scenes throughout physics' descent into the pictureless realm of the quantum. The real answer, if there is one, remained unclear. As we have seen, David Bohm has an answer, but it is startling. Matter is *not* stable, he says. It's only *relatively* stable and separate, relative to the flowing holographic senses of only relatively stable and separate observers. For Bohm, our universe is filled with nothing or no-things—it is a vast fluid no-thingness in which everything is.

In discovering his universe, Bohm has also discovered a surprising relationship between maps and terrains. He has discovered that for the looking-glass there can be no final maps, for our maps *are* the looking-glass. Our mapmaking changes the very terrain, and the terrain in turn changes our map. Maps, mapmakers, and terrains whirl around each other like the vortex in a river which expresses the whole.

Scientists and (flowing) history will decide whether Bohm's strange looking-glass insights will be retained.

WHOLE PARADOXES

For the moment, however, we might step back and appreciate that Bohm's implicate order resolves an amazing number of paradoxes and dualities that have plagued modern science and ancient philosophy. As with his fishtank analogy, apparently separate but correlated contraries are brought together here by moving to a higher dimension, another reality, from which the viewer can look back and see that what seemed two things was really only one all along. Here's a brief and incomplete résumé of this formidable success:

Particle vs. wave. The duality and paradox which kicked off the quantum revolution becomes, in the Bohmian world, simply a manifestation of the observer and his instruments intersecting in different ways with unfolding implicate ensembles.

Continuity vs. discontinuity. A problem going back to Zeno, it became a crisis with the introduction of the discontinuous quantum reality. In Bohm's implicate order, some ensembles (recall the dye drop) unfold into the explicate, giving the appearance of continuity; others unfold differently and appear discontinuous. Unfolding itself is both continuous and discontinuous at the same time.

Causality vs. noncausality. Bohm resolves this apparent paradox by taking causality to the furthest possible extreme. Everything causes everything else. What happens anywhere affects what happens everywhere. In our narrower dimension of space-time, objects and events (such as the appearance of malaria in human beings) can sometimes be abstracted and connected by small causal chains, and this is useful for practical purposes (creating vaccine).

Locality vs. nonlocality. Because everything causes everything else, a local event in the explicate order has its nonlocal root in the implicate holomovement.

Order vs. randomness. This duality ceases in a universe where what appears random at one level is ordered when seen from a higher level. A brother and sister to this dilemma is:

Determinate vs. indeterminate. Bohm provides a brilliant resolution to this conflict. The Bohmian universe is both determinate and indeterminate. Everything is ordered and determined in the holomovement. The indeterminate aspects of quantum mechanics have beneath them hidden variables which govern the outcomes of apparently probabilistic experiments. But these hidden variables also have hidden variables beneath them, and in principle these layers upon layers upon layers have no end and can never be finally determined because they aren't layers at all, they're holomovement. So though the universe is in some sense determined, it is in another more important sense multidimensional, creative, and indeterminate, constantly unfolding new "forms" or subtotalities which are expressions of the whole.

Known vs. unknown. Similarly, the known and the unknown are the same movement. Knowledge unfolds from the unknown and then enfolds back again. Consider, for example, how many unknown realities and unanswered questions our scientific knowledge has created, and how many things past ages were sure of that are now uncertain.

Creativity vs. stability. This is also an old question. It has two edges. One edge involves the surprising fact that forms in the universe remain stable for such long periods of time. Galaxies and solar systems endure for millions of years, species reproduce in the same form generation after generation. How is this possible? The other edge is: How can anything really new be created? Shouldn't reality be just repetitions and cycles taking place according to the fixed laws of nature? Bohm confronts this double-edged universe with a theory in which there are no fixed laws, only laws of relative stability. The laws of nature themselves evolve. Thus, the movement of the whole is the sheath which encases both edges of the paradox. Like the ocean of moving currents, it is the holomovement which keeps things stable. It is also the holomovement which from time to time creates new forms.

Being vs. becoming. Bohm's is a science of becoming. In the holomovement, being *is* becoming. Even as things stay the same, they unfold.

Micro vs. macro. Quantum mechanics introduced a split in the laws governing the micro and macro worlds, expressed by Schrödinger's cat problem (i.e., cats in many states can exist simultaneously in the quantum world but not in macro, human, reality). In the implicate order, the different laws describing micro and macro events are really only different abstractions of the whole, reliable in one domain, not in others. What we call the micro and macro worlds are different theoretical abstractions of the same world: the enfolding and unfolding implicate-explicate order.

Appearance vs. reality. Another brother and sister pair. Newtonian physics said the world is as it appears. Quantum mechanics said the world is at bottom different from the way it appears. Bohm says both. The resolution to this paradox is the same as above.

Part vs. whole. An old adage goes: The whole is greater than the sum of its parts. In scientific circles this has been a problem. The entire thrust of science has been, in effect, to explain the whole as the *sum* of its parts and nothing more. In the past, it was assumed that if some holistic principle were found, it could merely be added to the parts already known, as an organizer. In other words, the holistic principle would be something like an administrator who runs a bureaucracy. Bohm's approach reverses all this. For him, what we call the part is enfolded in and enfolds the whole; parts are only relative parts. They are, in reality, our abstractions (which are also enfolded in the whole).

Picture vs. no picture. Remember one of the complaints about quantum mechanics was that it provided no picture of what things look like or how they happen on the quantum level. Through his three models, Bohm does provide a picture of these workings, and beyond. But these are only very limited analogies. The holomovement itself is unvisualizable. What Bohm has provided is an "intuitive" picture, perhaps in something like the sense that the metaphor "Bring me my bow of burning gold" provides an intuitive picture of the speaker's passion.

Observer vs. observed. A very old dichotomy and one which lies at the foundations of experimental science. By assuming objectivity, scientists and their engineering colleagues have been able to reshape the world. So quantum mechanics came as a shock. Bohm comes as a greater shock. He says that the observer can do experiments with the observed, but he also *is* those experiments. He is also the observed, the face in the looking-glass. He can make abstractions about nature, but he is an abstraction too. As observers we are implicated in everything, birds, plants, particles; we are wound into the ecology and each other; to injure or affect anything is to affect our selves. The implicate and explicate order are one movement.

Brain vs. mind, mind vs. body. Matter and mind enfold each other. Brain is not separate from mind; on the other hand, the mind is not limited to the brain. Both are rooted in the implicate order, which is everywhere. "What happens in our own consciousness and what happens in nature are not fundamentally different in form. Therefore thought and matter have a great similarity of order."[22]

Time vs. timeless. Time is enfolded in the whole. Therefore time (or various time orders) are aspects of the timeless holomovement.

Besides these dilemmas, the implicate order harmonizes relativity and quantum mechanics by jumping to another dimension and looking back to see these theories as limited descriptions of a reality that lies beyond them. Bohm's theory finds a way out of the measurement problem and the nonlocality problem, the two closely related issues possibly fomenting a quantum mechanical crisis.

He has also dealt with such observations and concerns as led to wild and seemingly absurd speculations like Everett's multiple universe and the effects of the collective consciousness on matter. But he has dealt with these concerns in a consistent and logical manner that answers hosts of other questions as well.

PARADIGM STRIFE

But there are other issues his wholeness theory does not resolve. The grand unification approach to finding a common equation for gravity and the other three forces (strong nuclear, weak nuclear, and electromagnetic) is not of immediate interest to Bohm, who maintains that such research is based on the illusion of parts. Other physicists reply that Bohm's theory is more metaphysics than physics, devoid of experimental evidence, and scientifically vague. Bohm claims that appreciating the implicate order will lead to new kinds of experiments and the discovery of new data (as happened with Dalton's predicted atomic weights).

The contention here recalls Kuhn's depiction of what happens during a scientific revolution. The new paradigm declares some lines of previous research irrelevant. Advocates of opposing paradigms talk through each other because they're looking at different worlds.

In Bohm's world such metaphysical issues as ethics, morality, and truth are as important to science as evidence. For without such concern the scientist falls prey to self-deception.

Though scientists may profess to have no philosophy of science beyond that of open inquiry, they deceive themselves, he says, for their whole approach is based on unquestioned assumptions as to how scientific research should proceed and received prejudices as to how the world works. In the past century, for example, scientists have generally been insisting that it is self-centered and childish to believe that the universe is any more than the chance collisions of the indifferent forces of nature. Concealed in this attitude, Bohm says, is a danger. Scientists believe they can discover and harness the indifferent natural forces through determining nature's laws. This causes researchers to deceive themselves about the fact that science's very participation in formulating these laws may change them, and more deeply, that there can be no ultimate fixed laws of nature because nature is creative.

Various philosophers have emphasized—or in Bohm's rheomode speech, "re-levated"—different aspects of the progress of science. Bacon and Mach drew attention to its reliance on empirical data, Descartes and Popper to the logical construction of its theories, Kuhn to the fact that most scientists tend to assume common attitudes or paradigms toward knowledge and maneuver within these limits until a revolution occurs. Bohm, consistent with his creative universe, wants to stress science's creative and artistic aspect. Many people believe that physics is exact and that all its statements have exact meaning. On the contrary, Bohm believes, science is really closer to art in that it deals with things that are ambiguous. He calls scientific theories "art forms" which are constructed to fit with general experience but which can never give us any complete security of knowledge.

He sometimes tells a story about Bohr and Einstein which illustrates his deep apprehension that scientists and the rest of us may forget that our ideas (including the idea of the implicate order) are limited.

We mentioned that Einstein had felt a very great affection for Bohr. Later, it is told, when the two fell into animosity, some students sensed the sadness of this rift and hoping for a reunion, invited them, unbeknownst to each, to a party. Despite this effort, the two men refused to speak to each other and all through the party remained at opposite ends of the room. For Bohm, this human tragedy was rooted in the mistaken identification the men had made with their ideas. They had become frozen in their theories. On other occasions, Bohm tells how he was himself a witness to Einstein's pain over having lent his name and prestige (though never his scientific skill) to a request to President Roosevelt that initiated the atomic-bomb project. Such vignettes have a point. They provide a glimpse into why Bohm insists that implicate-order science must be seen as activity appropriate to the whole. The scientist can't hide behind the standard of objectivity and valueless research. The observer is the observed, the scientist is the looking-glass. But Bohm is evidently not advocating some easy solution, such as scientists' refusing to work on military projects. He is calling for a complete shift in the way scientists look at the universe and their work as an expression of the universe.

It might be tempting to dismiss these aspects of Bohm's thesis and concentrate on research that would validate and make use of the implicate order. Newton eventually abandoned physics for religious pursuits, saying science was too narrow. In the way in which he has formulated his theory, Bohm has tried to make it harder for spiritual, moral, humanistic aspects to be eliminated as if they were separate issues. But he knows the habit of fragmentation runs deep.

When listeners to Bohm draw parallels between his insights and those of philosophers like Plato, Leibniz, or Hegel or, as happens more frequently, between his view and that of Eastern metaphysics, Bohm replies that such parallels might be valid but that he prefers to think the issues through freshly from the beginning, so as to make each thinking process whole. It is easy to miss the importance of this point. First, Bohm is being consistent with his own theory of the holo-movement. Drawing parallels, comparing one system of thought to another, is a fragmentary approach and blocks the perception of the whole attainable by carefully unwinding a single thread.

Second, it is the common way human beings have of avoiding the shock and apparent danger of a mystery. We try to make the unfamiliar familiar by comparing it to old ideas rather than letting the old ideas drop and seeing what will happen. It seems we're afraid that if we do this we will fall through the mirror. So we try to look at the image from just one side. Too late, Bohm says, because, like it or not, recognize it or not, whatever side we're on, we have fallen through.

In this chapter we've dealt only sketchily with Bohm's scientific perspective, with the art form of his theory. But there are other—as Bohr might say—"complementary" Bohms. There are his considerations on the nature of truth and reality, thought and altered states of consciousness, religion, mysticism, social order, and death. Each of these considerations is subtle in itself and adds further depth and richness to the picture of his implicate order. It is doubtless impossible ever to sum up any human being's deep reflections on the mystery of life. That is especially the case with someone like Bohm whose reflections on the big questions are so intense. His thought is full of sudden clarities and unexplained obscurities, and these sometimes go together in a whole which is as much beyond him as it is his listeners.

Though he remains somewhat at odds with his colleagues in physics, Bohm's theory has already given considerable impetus to one area of research: the paranormal. The reasons are obvious. The idea of a nonmanifest nonlocal implicate order lying beyond things, an order which can be investigated (made explicit), seems made to "order" for the study of ESP and psychic realities.* Bohm himself has even participated in some experiments with famed psychic Uri Geller. Nevertheless, he argues that such research runs the same risk of self-deception as pursuit of the grand unification theory. It is easy to forget that the theory and the results are perceptions, not real separate parts or actual truths. Holistic theories and practices—from holistic health care to the holism of Eastern

*See Appendix of Related Expeditions, page 275.

religions—receive the same caution, as does the popular and ancient idea of the "spirit" or "soul." Such approaches do not receive Bohm's blessing because they are often ill-thought-out and ultimately fragmentary in their approach, despite claims to wholeness.

Whether Bohm's universe is "true" or will be accepted as such, it stands as a remarkable feat of fused insight, imagination, and logic. One can't help but be astonished at the degree to which he has been able to break out of the tight molds of scientific conditioning and stand alone with a completely new and literally vast idea, one which has both internal consistency and the logical power to explain widely diverging phenomena of physical experience from an entirely unexpected point of view. At the same time, the theory remains provocative, creative, and open—extending off into mists and depths still to be explored. It is a theory which is so intuitively satisfying that many people have felt that if the universe is not the way Bohm describes it, it ought to be. Nature, of course, has a way of being unappreciative of grand ideas.

Though many theoretical physicists at the moment reject Bohm's theory, their respect for him as a thinker is evident, and there has been a detectible swing to ideas of wholeness which may at least in part be attributable to his lifetime effort to give wholeness a voice.

In the next three sections we'll explore three theories originating from different branches of science which also give wholeness a voice, or different voices—perhaps the first sounds of a choir. They, like Bohm, provide maps to (but not of) the looking-glass.

3

The Sudden Vortex of Ilya Prigogine

You don't know how to deal with
Looking-glass cakes. Hand it round
first, and cut it afterwards. . . .

9. Entropy and the Paradox of Life

The second excursion to the inside of the looking-glass has its starting place in a different branch of physics from the solid-state and particle physics which were David Bohm's point of departure. It will extend through the territories of chemistry and biology. On this new journey we'll meet again the looking-glass principles introduced to us by Bohm—such principles as the observer is the observed, everything affects everything else, wholeness as flowing movement, the inanimate world as alive, a multidimensional reality, and the changeableness of the laws of nature. But we'll meet them from a different direction, in different language, and they will once again seem strange. Their strangeness may provide new insight into how Bohm's implicate order becomes explicate—how and why in a smooth-flowing river a vortex suddenly appears.

The nineteenth-century German poet-philosopher-scientist Wolfgang von Goethe objected to the scientific revolution of his day because he said it didn't explain "becoming" in nature. Science was too engrossed in explaining cause-and-effect relationships on the surface of things and missed the dynamic creative activity beneath. As we've seen, Bohm would agree. Nobel Prize–winning theoretical chemist Ilya Prigogine would also agree. He, like Bohm, has invented a "science of becoming."

THE PHYSICS OF DYING

Prigogine is a Russian-born Belgian who has spent his life studying one of the great nineteenth-century contributions to science, thermodynamics—a theory which embraces both physics and chemistry and is as majestic in its structure as Newton's mechanics.

Most readers will be passingly familiar with a major conclusion of thermodynamics, the conclusion that derives from its concept of entropy—the idea that "the universe is running down."

The entropy concept emerged from the science of the industrial revolution. From about the middle of the eighteenth-century through the nineteenth, engineers and physicists studied engines and learned about the relationship of heat and work. One of the things they learned was that work and heat are interchangeable. If you apply work to two sticks by rubbing them together, friction (heat) is produced. Supply heat to a steam-engine boiler and a piston moves (work). Eventually scientists realized that all different forms of energy—mechanical, electrical, chemical, thermal—are convertible into each other. But there was an important restriction. Scientists learned that when they went from one form of energy to another, not all the energy made the conversion. It wasn't lost, because the total energy within a system remains constant; rather, some of the energy stayed in a form where it could no longer be used. No matter how cleverly an engineer might design an engine, not all of the energy from the heat supply could be converted into work. The engine would never operate at 100 percent efficiency.

Scientists discovered that this barrier to the free exchange of energy was the key to why perpetual motion machines are impossible. Suppose the work from a machine is used to produce heat and this heat is used, in turn, to power the machine. The new science of thermodynamics showed engineers that in each cycle some of the energy would be converted into an unusable form and, without an independent input of power, the machine would quickly run down. Thermodynamics related this problem of the running down of all machines to "the law of increasing entropy."

What is entropy? To understand this concept we must first consider the difference between "reversible" and "irreversible" processes. A reversible process is one in which a system goes through a change and is able to return, thermodynamically, to the exact point where it started. Examples of almost reversible processes are the gradual stretching and return of a spring, or the very slow compression and release of gas in a piston. During these near-reversible processes no significant energy becomes tied up in a permanently unavailable form. Thermodynamically the system begins and ends its cycle at almost exactly the same point and no appreciable change in entropy occurs.

What about irreversible processes? A dramatic example of an irreversible cycle is the internal combustion engine. When the mixture of gasoline and air is ignited by the spark plug it explodes violently, producing a shock wave that forces the piston down. Even though the piston may return to its original position at the end of the cycle the system itself is thermodynamically in a very different state. The gases, for example, have become involved in a violent chemical reaction to produce carbon monoxide and water.

In this particular example only a small percentage of the chemical energy available in the gasoline is converted into useful work, the remainder is lost as heat, mechanical vibrations, energy tied up in the exhaust gases, and so on. In this extreme example, the system begins in one thermodynamic state and ends up in a very different one with a large increase in entropy. Even very gentle processes are, to some extent, irreversible because some of the available energy is lost through friction, vibrations, and heating.

The ideal reversible processes are totally gentle, free from shocks, sudden movement, friction, and violent flows of energy. Irreversible processes are just the opposite. Their sudden changes, shocks, turbulence, and explosiveness act to disturb the correlations between each elementary part of the system. Irreversible processes are therefore always accompanied by an increasing disorder and this is exactly what scientists mean by entropy. As the correlations and order of a system are broken down the system's entropy or disorder increases and energy that is available for useful work diminishes.

So irreversible systems are humpty-dumpty creatures. They move in a particular direction and thermodynamically can't be put back together again. They can't be put back because of the law of increasing entropy. Reversible systems are idealized, but in reality irreversible ones are the rule and increasing entropy is everywhere. It's not just the gasoline in the engine which reacts chemically and can't be put back together. The piston itself is subject to mechanical shocks and eventually wears out. Metal loses its strength, it rusts; wood rots; planets slowly spiral toward the sun. All of these phenomena experience entropy increase—increase in the amount of energy which becomes unavailable during energy conversions.

The fact of entropy increase gives a direction to time. Time flows from past to future, from the day the car was new, its pistons shiny, to the day it's towed off to the junkheap. An engine may be repaired, pistons reground, but only at the cost of an increase in entropy to the machinery making the repairs—and that only puts off the inevitable. In our way, we are no different from the car. We also wear out. In the dimension of space it is perfectly possible to move in any direction by exerting a force, yet no force will change our passage through time. It is impossible to grow younger. Time has a one-way arrow, and thermodynamics shows us how to tell its direction. The arrow of time points toward an increase in entropy.

Thermodynamics came as a surprise to Newtonian physicists. You may wonder why, since the fact that things decay and lose their form over time seems common sense. The reason is that in Newtonian mechanics, time is totally *reversible* and has no arrow. In formal terms, the equations of Newton's mechanics have what is called "time-reversal symmetry." Corresponding to each solution there is an equally valid one in which the direction of time is reversed. An example to illustrate this point:

Suppose the collision between two billiard balls is filmed. This collision will look perfectly natural whether the film is then projected normally or in reverse. Both states of affairs obey Newton's laws, and as far as the motion of elementary matter is concerned, time has no unique direction. In this mechanistic, objective world of matter, the flow of time appears quite different from our common-sense impressions.

Of course, there are processes in Newtonian physics which appear to move in one direction. A cannonball fired from a cannon rises in the air, then falls and makes a crater. One does not normally observe cannonballs rising by themselves from the ground, gaining speed, and entering the barrel of a gun. But that's because this example involves a series of quite complex phenomena—air resistance on the cannonball, the effects of gravity, the dissipation of heat on impact with the ground, and the distribution of earth thrown up from the crater. Nevertheless, each individual process in itself is perfectly reversible and if they could all be individually coordinated, there would be nothing to prevent these events from taking place this way. It is just astronomically improbable that such a coordinated state of affairs could happen.

If a stone is thrown into a pond, the disturbance of the

water spreads out in ripples until it reaches the edge of the pond. According to thermodynamics, this is a clear example of how entropy increases and energy is dissipated—demonstrating the arrow of time moving from past to future. In Newtonian physics, however, all the motions of the individual water molecules are reversible and there is nothing to stop billions and billions of tiny disturbances at the edge of the pond from all coordinating in such a way that a ripple grows and moves inward to focus at the center of the pond. Such a process is not impossible according to Newton's reversible laws of motion, it is simply absurdly improbable.

The arrow of time, which in thermodynamics is so naturally connected to an increase in entropy, is in Newtonian mechanics tied up with the idea of probability and an increasing lack of correlation within a system. Is the flow of time then connected to the absence of improbable events and the evolution of probable ones?

The answer was given by Ludwig Eduard Boltzmann, one of the intellectual giants of the late nineteenth century. Boltzmann was a Viennese physicist who argued that the results of thermodynamics could be based on Newtonian mechanics provided that one was willing to assume the existence of atoms and molecules. In the 1870s an atomic theory of matter was highly controversial, but Boltzmann pressed on to derive the basic laws and equations of classical thermodynamics in terms of the motions of billions of individual molecules.

It was, of course, impossible to treat the motions of such a large number of molecules individually, so Boltzmann adopted the same tactics that an insurance company does when it computes life expectancies and the chances of particular accidents among a population of millions. Boltzmann assumed that the individual molecules obeyed Newton's reversible laws of matter but that there were so many collisions taking place between these molecules each second that their individual paths were essentially random ones. He then went on to calculate average values using the theory of probability. His approach was called statistical mechanics.

Boltzmann also showed that heat is molecular motion. In a hot gas the individual molecules are in more violent motion than in a cold one. If a hot substance is placed in contact with a cold one, then the faster molecules transfer some of their energy to the slower ones by means of collisions—heat therefore travels from the hotter to the cooler body.

In addition to giving a molecular description of heat, Boltzmann was able to explain the nature of entropy—that is, he was able to explain why some energy becomes unavailable during energy conversions, why this unavailable energy is always increasing, and just what "unavailable" means. Boltzmann said that *entropy is nothing more than molecular chaos* and that when things are left to themselves the most chaotic situation will eventually prevail.

The example of a pack of cards may help explain what this means. The pack will be first arranged according to value and suit. Suppose that the cards in the pack are interchanged by shuffling. Each interchange is perfectly reversible, just as the motions in Newton's mechanics are reversible, and any particular arrangement is as probable as any other. For example, one interchange may cause the ace of hearts to change places with the king of spades. The next interchange may reverse this arrangement and place the king and ace back in their original position. But equally probable is an interchange of the king for the two of clubs or the queen of diamonds or any other card. In fact, at each interchange there are an enormous number of possibilities and only one of them will result in the pack's having its original order.

In this way, one can begin with a perfectly ordered situation, carry out a set of random exchanges, each of which is reversible, and end up with a gradual reduction of order. The reason is that an ordered state is as probable as any other arrangement, yet the total number of disordered arrangements is overwhelmingly larger. The end result of interchanging a pack of cards is to run through an astronomical number of different arrangements, only one of which is ordered in suits and values. In Boltzmann's terms the entropy of the pack of cards was zero at the beginning and then increased to its maximum value: chaos. Boltzmann's predecessor Rudolf Clausius had been able to sum up the whole of thermodynamics in one sentence: *"Die Energie der Welt ist Konstant; die Entropie der Welt strebt einem Maximum zu"*—"The energy of the world is constant; the entropy of the world endeavors to increase." According to Boltzmann's statistical mechanics, order and structure must always give way to disorder and chaos. Energy becomes unavailable because the individual correlations within a system are destroyed or mixed up. This led physicists to conclude that the ultimate fate of the universe is a "heat death" in which galaxies and solar systems become a homogeneous, thingless soup of particles and atoms. As the molecules bounce around becoming more and more random, useful ordered energy degrades toward what scientists call "equilibrium." An understanding of the term equilibrium is important for the arguments that follow. *Equilibrium is the maximum state of entropy. In equilibrium, useful energy has dissipated to the sporadic jerks and kicks of Brownian motion.*

Fresh from the perplexities of quantum theory and David Bohm's vast holistic universe, it may seem at this point that we have fallen back into an older, narrower notion of order. From the perspective of Bohm's implicate theory, for example, the thermodynamical concept of entropy and the idea of an exchange of "heat" and "work" would be seen as useful but limited notions: As we observers see it, the randomized particles of entropy aren't available for doing work, Bohm

Imagine that the molecules in the fanciful chesspiece flask on the left are a heated gas. The excited organization of this heat will dissipate as the particles mix with molecules in the cooler flask on the right. In time, a state of equilibrium will be reached where the molecules in the two flasks will be totally mixed up and random.

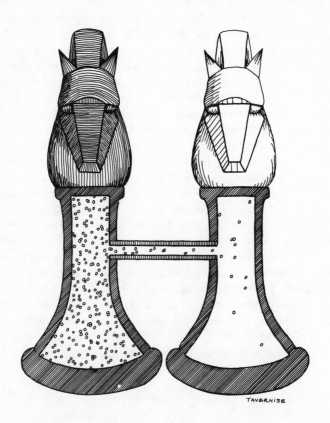

TAVERNISE

might say, but does this mean those particles are truly disordered? No, their order, like that of Brownian motion, is of a very high degree—and perhaps the "work" of these particles lies beyond the current explicate level.

Ilya Prigogine and Erich Jantsch (a scientist who applied Prigogine's theory to evolution) have, by another methodology, achieved conclusions very similar to Bohm's. In trying to merge quantum theory with the classical thermodynamics of Boltzmann, Prigogine found a separate entrance into the looking-glass. Perhaps this should not be surprising. Heraclitus, who said, "Everything flows," is also reported to have said, "You cannot step into the same river twice." Nor perhaps can you view the river of wholeness twice in exactly the same way. The path by which Ilya Prigogine arrived at his vision of the river lay first through a paradox that attended the very birth of thermodynamic theory.

A PHYSICS OF LIFE?

In the same era that Boltzmann and Clausius were developing their theory of universal entropy—the breakdown of forms as they progress toward disorder and equilibrium—Charles Darwin and Alfred Russel Wallace were formulating and promulgating a theory which considered just the opposite: the emergence of the high organization of *living* forms. The theory of evolution today describes a process in which individual atoms and molecules are organized into amino acids and proteins. More curiously, it describes the development of cells and increasingly complex organisms which, once formed, carefully readjust themselves to preserve their integrity for future growth. How can this be? How can life appear, sustain itself, and display organizational growth in the face of the universal march of entropy? Is life simply a chance process? Or is the universe in some way growing and evolving despite its apparent tendency toward randomization? Must the laws of thermodynamics themselves be widened to include the emergence of new and novel systems? These were the sorts of questions asked by Erwin Schrödinger and by Théophile de Donder at the Free University of Brussels. Working in the first decades of this century, these men sought to enlarge Boltzmann's approach to include living systems.

It was de Donder's spiritual successor at Brussels, Ilya Prigogine, who finally found an answer. Prigogine is a brusque man, in appearance more like a businessman than a scientist. His air of precision is mingled with a passion for collecting art works which show figures just emerging out of a blurred, amorphous background. Such art mirrors the theory which Prigogine proposes as a startling answer to the paradox of life.

In the nineteenth century, thermodynamics had been dominated by the machine and the engine, systems which had to be pushed toward maximum efficiency. As a consequence, physicists tended to place emphasis on investigating closed systems and almost reversible reactions in which the increase in entropy is minimized as a system moves fairly gently and hovers never very far from equilibrium. The old-fashioned steam engine is a good example, because its thermodynamic change is more gradual than the automobile; at the end of each cycle the system returns to a similar thermodynamic state. On the one hand, the steam-engine system is maintained far enough from equilibrium so that there is always enough temperature difference between the hot (steam) and cool (condenser) parts. If this difference were not maintained (as would happen if the condenser failed), the efficiency of the engine would drop as hot and cool parts headed toward equilibrium. On the other hand, the engine is not allowed to get very far away from equilibrium. If that were to happen—if, for example, the piston were to move in a jerky or explosive fashion—entropy-producing irreversible processes would dominate and efficiency would be low. To get maximum efficiency out of the steam engine as a whole, the system has to be kept on a tightrope, close to equilibrium but not too close.

The steam engine is what scientist call a "closed system." In a closed system there are clearly defined exchanges between heat sources and work—as when the piston is moved by expansion of hot steam in the cylinder and the steam is then condensed and fed back to the boiler. Nothing new enters or leaves a closed system. The system has clearly defined parts. For maximum efficiency these parts must keep to a fixed regime. They can only operate within very narrow ranges. A part can be replaced, of course, but the system itself doesn't make such repairs. That help has to come from the outside— usually in the form of a mechanic. Other kinds of closed systems are a rock and cool bowl of soup—items that have actually reached equilibrium. Thus closed systems always involve equilibrium or near-equilibrium situations.

It is not surprising that living systems were left out of this thermodynamic picture. In contrast to machines, life forms are "open systems" which emerge and actually thrive in a volatile arena *far* from equilibrium. Open systems are able to adjust to outside changes; they take in food, grow, replace their own parts, reproduce, and even survive the total loss of some parts—all without the aid of a mechanic.

Prigogine realized that a major factor distinguishing living, open systems from closed systems such as machines is the far-from-equilibrium environment—an environment of high energy and the influx of new chemicals. It is an environment which is "chaotic" in a different way from the thingless-soup chaos of entropy and equilibrium. It is, instead, chaotic with competing energy and chemical fluctuations. In this environment, living systems appear and evolve.

Prigogine decided to study the dynamics of the far-from-equilibrium state to see if he could discover there any clues as to how life performs this magic trick. He quickly learned it isn't only life forms which occur in far-from-equilibrium situations. These situations give birth to sudden orders of all kinds.

Turn on your tap and you can experience an instant example of the dynamics of an open system. If you open the tap slightly, you create a situation in which the water appears clear, round, and smooth. The water molecules in this flow are following a random, statistical law. If you open the tap farther, increasing the water pressure, the water first appears turbulent. This turbulence, however, quickly establishes itself as a stable *order*. Notice there are muscular strands. These are grouping the water molecules together into powerful streams which have the effect of allowing more water to pass through. This structure or order isn't solid, of course; it isn't made up of any static components. It's a structure of flowing. If you open the tap even farther, still another such stable flowing structure will form.

From sand dunes to the formation of stars and galaxies, in open systems stable flowing structures appear out of fluctuating energy flows. The plasma vibrations which Bohm studied and felt were "alive" are a subatomic example. Superconductivity is another: In many metals, at low temperatures

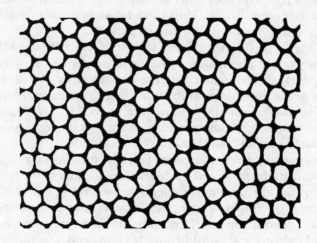

At certain critical temperatures when water is heated, convection currents appear and organize into a stable structure. Seen from above these appear as a pattern of hexagonal cells.

the electrical resistance indicates that traveling electrons are scattering, at random, off vibrating nuclei in the metal lattice. Yet at a certain critical low temperature all resistance vanishes as these chaotic motions expose an underlying order in which electrons behave in a collective fashion and the "superconducting" state appears.

At a social biological level, a termite's nest can also be understood as a far-from-equilibrium structure. Initially, termites make unordered deposits of matter. The random fluctuation of large numbers of them, however, eventually increases until it sets off a phase of highly organized activity which gives birth to the termite nest.

In 1958, two Russian researchers stumbled upon a far-from-equilibrium structure occurring in a chemical environment. When they mixed malonic acid, bromate, and cerium ions in a shallow dish of sulfuric acid at certain critical temperatures, what is now called the Belousov-Zhabotinsky reaction created a structure of concentric and spiral "cells" that pulsed and remained stable even as the reaction secreted more cells. The reaction is clearly chemical and does not involve DNA, but in its structure it looks like the growth of a life form!

The discovery of far-from-equilibrium structures appearing in chemical reactions offered new insights into the dynamics of these spontaneous forms. They are quite different from "regular" reactions.

When a spoonful of instant coffee is put into a cup of hot water, the coffee dissolves and spreads out randomly until it is uniformly distributed in the water. The total entropy of the system increases. Putting coffee in water results in a mixture where water and "coffee" molecules intermingle but don't bond to each other. In simple chemical reactions (which we are calling "regular" reactions) this process has the added dimension of bonding. If there are two chemicals, hydrogen and oxygen, for example, the two sets of "reagent" molecules undergo random molecular collisions until they are close enough and energetic enough to form bonds among each other and produce water molecules. The bonding itself follows quantum mechanical rules which govern the exchange of electrons, but the mixing diffusion and collision that precedes bonding takes place at random like the coffee mixing

The Belousov-Zhabotinsky reaction. A chemical reaction as a life form?

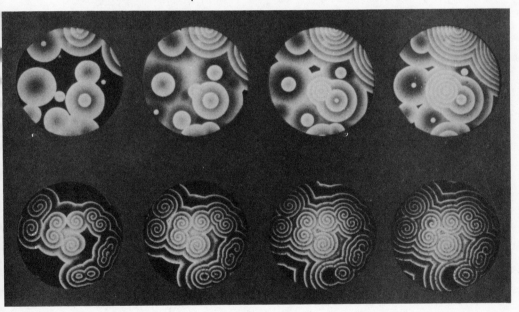

with water. As the reaction progresses, more and more of the two sets of reagent molecules bond together, the reagents are used up, and the new "product" concentration increases. Eventually all the reagents have been converted into the product and the reaction ends. Depending on conditions, the speed of this process can vary from many hours to a few seconds or even explosively.

In far-from-equilibrium chemical situations, something very different takes place. A specific concentration of a particular reagent or a given high temperature sets off giant oscillations in the reaction system. One of the reagent chemicals for example builds up in concentration, dies away, then

A diagram of the Belousov-Zhabotinsky reaction. The autocatalytic step is the circle marked A. X and Y are starting chemicals taken into the system. A reacts to form B, which in turn forms C, and this produces more of A. P and Q are end products of the reaction. The autocatalytic process ABC renews itself continuously and acts as a catalyst which transforms products (X and Y) into end products (P and Q). This process is typical of spontaneous reactions that appear in far-from-equilibrium situations, including reactions that take place inside living cells.

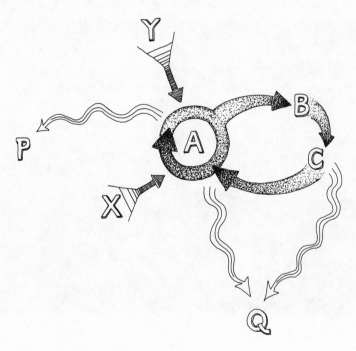

builds up again. Unlike the "regular" chemical reaction in which the reagents and products are distributed at random throughout the solution, here local inhomogeneities, or differences, form. In one region a particular chemical may dominate the reaction while in another its concentration is exhausted. Out of this chaotic flux, the reaction begins generating a stable structure in space and time.

A major feature of such chemical structures is the establishment of what is called "autocatalysis." Autocatalysis means that one of the products of the reaction enters a cycle that helps it to reproduce itself! The whole reaction pivots on the autocatalytic cycle. The cycle stabilizes the structure and maintains its continued shape and "flow."

The word "autocatalysis" comes from "auto" or self and "catalyst," an agent which aids in the production of other things but is not itself changed. Living cells are, of course, autocatalytic in the sense that they can produce more of themselves at the same time they preserve themselves in a changing environment. Thus the "nonliving" activity that often takes place within cells (far-from-equilibrium chemical reactions) has a structure similar to the living cell itself. Such a reaction is not simply a "part" of the cell's struggle for life. Such a reaction is its own life. The cells and the reaction, life and nonlife, are mirrors of each other.

David Bohm argued that the appearance of life wasn't a chance conjunction of molecules but unfolded inevitably from the multidimensional order of the universe. He insists that the usual scientific distinctions between life and nonlife are only abstractions, possessing limited value. In fact, life and nonlife are implicately woven into each other. Traveling along on his far-from-equilibrium expedition, Ilya Prigogine encountered a similiar discovery. Life and nonlife both appear in nonequilibrium situations—and such situations are everywhere. Nineteenth-century thermodynamics had portrayed a universe in which entropy increases and structures inevitably break down. Prigogine discovered a thermodynamics which describes how in far-from-equilibrium situations structures will inevitably form. He called the dynamics of such structures "order through fluctuation."

At first, far away from equilibrium, the fluctuating motion of an individual molecule or atom or termite is unpredictable. Then a critical point is reached and the random motions change into fluctuations of larger numbers of molecules and, in turn, these fluctuations suddenly increase in size. Prigogine says the fluctuations are "amplified" until they no longer average out. The system (such as the electron plasma or the chemicals in the Belousov-Zhabotinsky reaction) is on a knife edge, "bifurcation" (branching) point. As the system approaches a critical point it is not going in any general direction but has the potential to move in any one of a number of different directions, and it's impossible to know which one will predominate and direct the system. Then suddenly one takes over and a new order is established which becomes highly resistant to further fluctuation. The new order is flowing yet stable for it can only change if the system is subjected to very intense further fluctuation. At some point, if this new fluctuation cannot be "damped" by the structure, the structure is broken out of its shape (as when you open the water tap further), a new phase of chaos occurs, and a new and "higher," more complex order evolves. Several writers have compared this evolution of complexity to how new social orders evolve out of political crisis, new psychological strengths out of suffering and conflict, new artistic forms out of the apparent chaos of an artist's creative process.

It can also be compared to Bohm's wholeness. For in fact, one of Bohm's images for wholeness, the vortex in a flowing river, is a far-from-equilibrium structure. And in this image the insights of the two theories merge.

Eddies and ripples in the river grow and die away at random while the river's flow is undisturbed on average. Then, at some region, the water becomes turbulent and uncertain; the slightest change in external conditions (such as someone throwing a rock into the water upstream, or an increase in the water's volume) may push these fluctuations in one direction or another. Such a fluid system is unstable and unpredictable. Without warning one direction takes over and a vortex is formed. It is an ordered system, simultaneously autonomous and inseparable from the river. Once formed, the vortex remains remarkably stable in the face of further fluctuations and change. Yet subjected to much greater perturbation, it may spontaneously disintegrate or evolve into a new form.

Discovering the dynamics of the vortex and other far-from-equilibrium structures was an early step. When Prigogine began to trace out the implications of his discovery, he found himself, like Bohm, spontaneously transported into a new and far-reaching universe.

10. Founding a Science of Spontaneous Order

Prigogine called far-from-equilibrium forms like the vortex, "dissipative structures." The name comes from the fact that to keep their shape these structures must constantly dissipate entropy so it won't build up inside the entity and "kill" it with equilibrium. To dissipate entropy requires a constant input of energy and new materials, which is why dissipative structures must form in energy-filled, far-from-equilibrium situations (a vortex wouldn't form in a still pond).

In classical thermodynamics, a measure of the efficiency of a process is a low rate of heat loss or entropy production. Dissipative structures are effective for precisely the opposite reason. In its high generation of entropy and continued openness to fluctuating energy input, the dissipative structure literally transcends closed-system thermodynamics. This also makes the dissipative structure a paradox. It can survive only by remaining open to a flowing matter and energy exchange with the environment. In fact, matter and energy literally flow through it and form it, like the river water through a vortex. On the other hand, this very openness somehow makes the structure resistant to change. The internal environment of a warm-blooded animal, for example, maintains a stable temperature and chemical constitution despite wide variations in the environment outside. This means the resistance to change must itself be a kind of flowing. The structure is *stabilized* by its flowing. It is stable but only relatively stable—relative to the constant energy flow required to maintain its shape. Its very stability is also paradoxically an *instability* because of its total dependence on its environment. The dissipative structure is autonomous (separate) but only relatively separate. It is a flow within a flow. At its "edges" is a constant exchange of one flow to another. More and more, it looks like a creature out of Bohm.

As Prigogine explored the implications of these strange mirror-world phenomena, he encountered the problems of wholeness. In the last chapter we mentioned that if a dissipative system is subjected to intense perturbation and is unable to damp these jolts and jars to its structure, it may, in Prigogine's words, "escape into a higher order." But there's another possibility (actually, as we'll learn in the next chapter, it's the same possibility). The system may generate inside itself on a "lower level" a dissipative structure which compensates for the change. An example of such internal compensation is a dissipative chemical reaction taking place inside a cell. If we stand inside the chemical environment of the cell and watch, this sudden dissipative form looks like an instance of spontaneous order appearing out of fluctuation. However, if we stand outside the cell and look at it, we get a very different perspective. Here it looks as if this reaction is an internal mechanism by which an established dissipative structure (the cell) is able to maintain its shape in the face of a change in its external (and consequently internal) environment. Putting these two perspectives together spontaneously generates a question.

Should it be said that the cell is sustaining the dissipative chemical reaction, or should it be said that the chemical reaction is sustaining the cell? It is a chicken-and-egg question perhaps, but one with profound implications.

CRACKING THE CHICKEN-EGG QUESTION

Traditionally, scientists have believed there is a hierarchy of explanations for the universe. On the most fundamental level is physics, the movement of the simplest parts of particles. Next comes chemistry, which describes how atoms combine to form molecules. After that comes molecular biology, from simple collections of macromolecules up in stages to biology and presumably the most complex organisms, human beings. The hierarchy leads finally to the very complex explanations of neurophysiology (the study of the brain and nervous system). Prigogine thinks assuming this hierarchy is a mistake. He argues that there is no real hierarchy, no fundamental level of description with other levels stacked on top of it. Instead there are *different* levels, each dependent on the others in complex ways. To describe the dissipative chemical reaction in the cell, we could look at it from the level of atoms and molecules or we could look at it from the level of the cell. "Lower" levels depend on "higher" levels for their existence as much as higher levels depend on lower ones for theirs. One level doesn't come "before" or "after" another in a hierarchy. The universe can't be disassembled into simpler and simpler parts. Everything is in dynamic interaction; it is a seamless, fluid web of process structures.

One senses here many similiarities to Bohm's implicate order. Bohm also insists no complete or fundamental description of nature is possible, and would agree that each level affects the others (in the holomovement). Bohm asks us to realize that the "laws of nature" can only be laws of a relatively autonomous subtotal, that is, laws of one level or limited domain. Beneath, beyond, and above each level there will always lie other levels in a multidimensional reality. And even the limited laws of levels are subject to evolution and change.

For Prigogine, too, reality is multidimensional. The laws of each level of reality are different from the laws of other levels. The laws of atoms, the laws of molecules, the laws of consciousness, the laws of sociology—all are different. No level is the fundamental level or foundation block on which the others can be hierarchically arranged. As evolution takes place and more complex dissipative structures (such as human society) appear, they bring with them new levels and new laws. So the laws of nature evolve as new dissipative structures evolve. Prigogine thinks that instead of arranging levels in a hierarchy (physics, chemistry, biology), we should begin a dialectic, a kind of animated discussion between levels to create, in effect, a web of description where higher levels feed back strands of information to lower levels, back to new higher levels, and so on—mirroring the web of reality with our theories. In this way, we will see that our description of the laws of nature will evolve as new levels of complexity evolve. This will create new levels of complexity in our theories. Prigogine's picture of unfolding reality is uncanny—not only in its similarities to Bohm's, but also in its differences.

Prigogine's approach takes him, inevitably, to the center of the looking-glass:

If the universe isn't constructed from the bottom up but is a web of levels and differing laws, where do we, as observers of this universe, stand? Obviously not objectively outside it, because our own processes are entangled in the strands. Here, again, Prigogine's conclusion is very like Bohm's, viewed from a totally new perspective. This is his logic:

Human beings live an irreversible existence. Our arrow of

time points toward death. The evolution of the dissipative structure we call consciousness brought to the history of the universe a new level of reality and a new law of nature. This new law involves the observer's ability to appreciate the difference between past and future. When the observer looks at the microscopic world of matter and energy, he observes that at that tiny level past and future (time) are reversible. But this observation of a reversible microscopic universe is made by a macroscopic being who is himself irreversible and knows it. The reversible universe (observed) conditions the irreversible existence (of the observer). In turn, the irreversible observer's existence conditions the whole meaning of observed reversibility, for without the observer there would be no idea of reversibility at all. So a circle is completed: The observer is the observed. Reversible and irreversible are levels infused in each other.

These are Prigogine's looking-glass realizations. How does he validate them in the languages and logics of science? He has taken as his major task to show that the irreversibility of the macro world of cats and consciousness is just as fundamental as the reversible micro world of small-particle physics. On another level, reversibility and irreversibility are concepts which embody Prigogine's fascination with *time*.

Since the days of Boltzmann, irreversibility has been accepted as a real or objective phenomenon on the biological or macro level. Remember that Boltzmann's statistical mechanics was based on a Newtonian model of molecules and atoms which envisioned them as time-reversible billiard balls careening about a table. Boltzmann resolved the irreversible-reversible paradox by showing how the reversible equations of classical physics could still fit the irreversible universe of entropy. All structures above the atomic level, he said, will eventually break down as the atomic cards are shuffled and equilibrium is reached. With the advent of quantum theory, deep confusion entered this picture and a new paradox between reversible and irreversible emerged. The equations of the Schrödinger wave function are also time-reversible, like those of classical mechanics. But the measurement process and "collapse of the wave function" is irreversible. Probability also features in quantum physics but the randomness of quantum processes is *not* like the shuffling of a deck of cards. Quantum probability is *given*. It cannot be resolved by assuming, as Boltzmann did, that in theory things could be reversed but that this reversing simply becomes increasingly improbable. That idea won't work here. In quantum theory the reversible equations and the irreversible measurements observers make are unresolvably at odds. This is apparent the moment the wave function collapses. The moment before that happens, the many solutions of the Schrödinger equation exist simultaneously. Time is reversible; the particle is spread all over space. A moment later there is a single click on a detector, observers see the particle is in one place, there is one direction for time, the cat is dead or alive. Between these two moments quantum theory offers no explanation.

Because quantum theory assumes that it is dealing with the smallest, most elementary parts of nature, the mathematics of the theory is generally considered to be the most fundamental law of nature. What we observers see on our level must be derivative from this law, scientists believe.

Prigogine doesn't agree. He has sought to resolve the Schrödinger cat paradox by giving the concept of irreversibility an equal footing with reversibility as a law of nature. He has tried to show in scientifically convincing ways that neither is more fundamental, there is no hierarchy, everything affects everything else, and the observer is entwined in the observed. To make his case, he has, like Bohm, turned to mathematical techniques that extend beyond previously accepted orders, primarily to two approaches. The first approach involves the concept of "nonlinear equations," the second "symmetry breaking."

Until the arrival of computer solutions and twentieth-century mathematical analysis, nonlinear differential equations were particularly difficult to solve, much more difficult than their cousins, linear equations.

Linear differential equations are special or simplified cases of the more general nonlinear equations. During the nineteenth century, mathematicians and physicists derived standard approaches to solving these linear equations. The mathematical "shape" of the equation would give considerable information about its solutions and enable them to be classified. Such equations did not hold surprises, and when one solution was determined the others would follow.

Suppose, for example, the linear differential equation represented two mechanical systems in interaction. As this interaction increased or decreased, the behavior of each system would change in a predictable manner. If the interaction increased by a very small amount, the behavior of the system would likewise change by a small amount. As the interaction decreased toward zero, the mechanical behavior would approach that of two independent systems. In linear systems, therefore, small changes produce small effects, determinism is everywhere apparent, and by reducing interactions to very small values the system can be considered to be composed of independent parts.

Not all systems are linear however—in fact, very few real ones are—but physicists could assume that, provided systems stayed very close to equilibrium, a linear approximation would be a good one. Because linear equations were so well understood and because linear systems behaved as if they could be broken apart into independent units, scientists throughout the nineteenth century grew increasingly confident about a linear world. Of course, there were problems that remained stubbornly nonlinear, but, wherever possible, mathematical physicists would attempt to "linearize" a system and treat the nonlinear parts as corrections.

The mathematical treatment of nonlinear differential equations is far more forbidding than that of linear ones. Their solutions are not always obvious or straightforward, and obtaining one solution may not be much help in obtaining others. Most important is how these solutions change when interaction terms are modified. In a linear system a small change in an interaction will produce a small change in a solution. This is not the case with a nonlinear differential equation. The solution may change slowly as a parameter in the equation is varied and then suddenly change to a totally new type of solution. This change of behavior is dramatic and unpredictable—for a whole range of values the system may behave in a regular and reasonable way but an additional, infinitesimal change of the same parameter may throw the system into a wholly new state. The effect is rather like the last straw which breaks the camel's back. Bales and bales of straw are added and with each bale the camel sags a little. Finally a single straw is placed on the top and the camel and its load crash to the ground. A formal mathematical expression of just such occurrences was given in the 1970s by the French mathematician René Thom. He calls his approach, appropriately, "catastrophe theory."

The linear world is a world without surprises. It is a

clockwork world in which things can be taken apart and rebuilt again. By contrast, the nonlinear world described mathematically by Thom can be violent and unpredictable. The mathematician cannot just set his equation in motion and let it grind toward an inevitable solution; in a metaphorical sense he must enter into it, affecting, judging, evaluating the situation as it develops. The observer is always present and cannot abstract himself from the system. Prigogine asserts that the nonlinear world contains most of what is important in nature. It is the world of the dissipative structure. One of Prigogine's major successes has been his ability to find ways of mathematically treating these irreversible unfolding structures.

To make his case that irreversible time must be given equal footing with reversible time as a law of nature, Prigogine marshals another mathematical idea.

Boltzmann had shown that the apparently irreversible march of entropy was a "coarse-graining" effect or statistical averaging of the continual shuffling of atoms. Time was an illusion produced by the statistics of large numbers of particles. Seen on a fine enough scale, time would be totally reversible.

In contrast, Prigogine thinks that while the laws of motion for molecules may be reversible, the true nature of time is irreversible. How can this be possible? The answer comes from the symmetry-breaking concept which originates in the mathematical side of quantum physics. Laws of nature express the possibilities or potentialities for matter. They set out boundaries for behavior and circumscribe what could happen. When these laws are expressed mathematically, the formulae show certain symmetries. This had led physicists to conceive of the universe as essentially symmetrical.

At first blush this seems to be quite an idealized concept. Physicists imagine that without matter, everywhere in the universe is exactly the same; any place in space is as good as any other. Provided there is no matter in the neighborhood, an experiment done at one point in space will yield the same results as one done at another point.

Reality, of course, is different. Stars and planets warp the curvature of space. Particles drift and flow through the vacuum in clouds. One point is not the same as any other.

Physicists say this is because the symmetry of space has been "broken." Imagine the concept this way: Suppose you were standing on a giant blank sphere (with suction shoes so you wouldn't fall off). Anywhere you walked on the sphere would be the same as anywhere else. There would be no direction. The sphere would be perfectly symmetrical. Now imagine you discovered a tiny crack beneath your feet. That crack in the ball would give a direction. It would break the symmetry. Physicists today expect symmetries to be broken. Whenever an event occurs or a structure appears, symmetries crack.

Another way to think about symmetries is to imagine nature as a roulette wheel. The wheel is whole and symmetrical and contains a number (in nature an infinite number) of slots for the ball to fall into. Each of these slots is a potential existence, each with an equal chance of occurring. Yet the ball will fall into only one hole. The fact that only one reality can emerge from a host of equal probabilities means that symmetries must always be broken for anything to occur.

Ilya Prigogine

Prigogine applies this concept to the nature of time. While the equations of nature are time-symmetrical, real processes are not. Dissipative structures break the symmetry of time the way the appearance of an electron breaks the symmetry of space. This symmetry still exists while the system fluctuates at the knife edge or bifurcation point—but once the dissipative structure forms, it goes in only one direction. Since dissipative structures are coupled together (think of all the dissipative structures coupled together in your body), they take a consistent direction which we, as macroscopic observers, call a direction from past to future.*

Irreversible time to Prigogine is therefore not an illusion produced by the coarse graining of matter, an averaging effect, any more than a particle is an illusion—or they are equally illusions. Rather, irreversible time results from the symmetry breaking of reversible time by real macroscopic processes. In his thermodynamic equations, Prigogine introduces what he calls a "time operator," T. This operator corresponds to "historical time," that is, an internal time or age of a system, expressing time's one-way flow. Here there is an interesting connection to Bohm's idea of time.

For Bohm there could be various time orders, each unfolding at a different rate, which would not necessarily be the same as the time order unfolded by a clock. For Prigogine, similarly, each dissipative structure has its own time order, T. The life span of a fly that survives only one human day has its own time. To compare it with a human life span is to make an error. Both are lifetimes, each unfolding in a different time, a different T. He says the time measured by a clock is a kind of average T established by us as observers.

Prigogine goes on to treat the reversible time, t, used by physicists in the equations of classical physics and quantum mechanics as only a parameter which has to do with the motion of particles. By making a distinction between T and t, Prigogine is able to point out that the Schrödinger cat problem which has so vexed physicists actually results from a confusion between historical time, our time (T), and the abstract motion of possibilities (t). The parameter t in the

---

*Readers may want to compare the idea of dissipative structures coupling together with Bohm's attempts to use the Lorentz time-space frames (page 142) to show how events couple together to create an explicate order. Remember the illustration of the wired-together clocks.

quantum mechanical equations describes the movement of the wave function as it "leaks" symmetrically out of the nucleus. However, the actual experiment which the scientist observes takes place in the broken symmetry of irreversible time T. For Prigogine there aren't any multiple cats in various stages of aliveness and deadness. He argues being alive or dead doesn't enter the picture until time T is introduced. Time T is introduced with a dissipative structure. In the Schrödinger cat experiment there are two dissipative structures, the cat and the experimenter opening the box. Each of them breaks the symmetry of time. Until that point scientists are not talking about real events but about symmetries in the laws of nature. In the broken symmetries of the real world, the cat must be either alive or dead.

For Prigogine, the laws of dissipative structure are as fundamental as any laws in nature. They are the laws that give shape not only to space but to time. They move the universe "from being to becoming." We could also put this in other ways: For the dissipative structure, being *is* becoming.* A dissipative structure doesn't emerge into time, it *is* time.

Despite the many deep similarities in the ways Bohm and Prigogine treat time, Prigogine's universe gives time a primary emphasis. The laws of time are his key to showing that no level of reality can be considered more fundamental than any other, that no hierarchy can exist. In awarding Prigogine the Nobel Prize for chemistry in 1977, the Nobel committee honored him for creating theories that bridged the gap between various sciences—that is, between various levels and realities in nature.

OTHER HATCHINGS

In the early 1970s a theoretical biologist from Chile, Humberto Maturana, developed the idea of what he called "autopoiesis" (from the Greek, meaning "self-production") as a way to describe living systems. Two Chilean colleagues, Francisco Varela and Ricardo Uribe, gave further expression to Maturana's notion. Their looking-glass journey reached

---

* So Prigogine agrees with Bohm who said that in the implicate-explicate order, being is becoming—for in that order being is continually unfolding.

Prigogine's world of dissipative structures by a different route.

The Chileans compare their "autopoietic systems" to "allopoietic systems." They say the allopoietic system, your family car, for example, contains the same molecules from showroom to junkyard (except, of course, for some spots of inevitable rust). The car also has no identity problem. Or rather its problem is that it has no problem because its identity is totally given to it from the outside. The car is only what you and Detroit (or Germany or Japan) say it is and how it is used, nothing more.

In contrast, autopoietic structures like the family dog or the plants in the window change molecules all the time and yet remain somehow "the same." Such structures are their own reason for being. A farmer may cultivate a field of wheat and think about it the way he would the automobile, as just what it is used for (flour, cereal); nevertheless, the identities of those golden stalks blowing in the wind are entirely independent of human definitions. Their identities are made by themselves internally. How? What is this self-contained identity?

Obviously it isn't only the physical shape of the wheat, which is always growing and changing. A person's definition of himself, for example, is not limited to his cells. Parts of the body could be paralyzed or amputated and the person would still be "the same person," still whole despite this loss of parts. Then what is that person?

The Chilean biologists say the identity of any living entity comes from its relationships with its environment. This may sound familiarly paradoxical. It's the dissipative-structure paradox dressed up in new language. A wheat stalk is autonomous (separate); its autonomy derives from its interdependence with its surroundings.

The key to resolving this paradox, these biologists say, is how we understand words like "process" and "relationship." We're so used to thinking of "things" that this may be difficult to grasp at first, although we've had some good practice in letting go of our belief in "things" with Bohm. Here it's important, too, because the whole dissipative-structure paradigm has replaced "things" with "process structures."

Dissipative or autopoietic process structures are not like gears always going around the same way. Take as an example an individual wheat plant. A wheat plant's identity is defined by an intricate web of connections with sun, air, soil. These connections involve complex molecular reactions which convert matter and energy from one form to another to maintain the dynamic balance that is the plant. This dynamic balance means all the various processes in the wheat stay in the same relationship to one another but are constantly moving. If the soil becomes slightly more acidic, the plant's chemical systems will shift over to compensate and everything else in the web (all the other chemical reactions we call the plant) will adjust accordingly. Put another way, even if a part is lost, identity will be retained—the web will fill in. Changes inside and outside can be adjusted because the wheat relationships aren't among separate parts. What we call "parts" are really different expressions of a whole movement.*

The sole purpose of this moving web is to continually reproduce itself (reproduce its structure). As all this happens, the web which is the wheat stalk gives a shape to space and time. Sexual reproduction or cell division is, in this light, merely a further aspect of the drive of autopoietic structures to continually reconstitute themselves.

Because the various processes in an autopoietic system always remain in the same dynamic relationship to one another, the system is described by Maturana, Varela, and Uribe as closed, "information-tight." Once again the paradox: On the one hand the autopoietic system is in constant exchange with the environment, on the other it is locked into its own tight order.

Erich Jantsch added a further twist to this paradox.

Jantsch was a scientist with an encyclopedic mind. Before his death in 1980 he was a major synthesizer of ideas circulating around Prigogine's dissipative-structure paradigm. He expanded on the autopoietic concept and added it to Prigogine's approach. Jantsch defines autopoiesis as the state of a dissipative structure once it has gone through the turmoils and turbulences of youth and adolescence and "established" its identity in the far-from-equilibrium environment. He calls those dissipative structures which become autopoietic "self-organizing structures."

Jantsch says self-organizing, autopoietic, structures keep

---

*Readers might want to consider the question of how theories of wholeness explain organ transplants. There is a discussion of this in Part 4, Chapter 14, page 223.

the shape of their processes by constantly balancing the need to remain safe from fluctuation with the need to remain open to it. He observes that in human beings, for example, "lower-level" processes like the circulatory system and the chemical activities of the digestive tract are fairly habitual and "closed" to fluctuation. They adjust constantly to maintain balance, and if a person eats something he shouldn't or gets stung by a bee, these systems try (if they can) to reject or damp the new input. The "higher-level" systems, like the brain, however, are more open to fluctuation. They also try to maintain a dynamic balance, but the brain is inherently more unstable. In the human brain, for example, thoughts bounce around and can produce great fluctuation (such as anxiety, hope, fantasy) before they're rejected. Some radically new thought may not be rejected at all but may generate so much fluctuation that it will transform the brain with a new insight or new sense of self. (Such transformations could even cause changes in lower level processes. For example, changes in the circulatory system and muscles might result from a new self-image that led to a regimen of diet and exercise). Jantsch, who was an organizational theorist as well as a scientist, notes a thought-provoking corollary to this insight into biological hierarchy: Biological organizations put innovation and creativity at the "top" of the hierarchy (in the brain), just the opposite from their position in most human-made organizations (as corporations or nations), where the structure becomes increasingly rigid toward the top.

Jantsch viewed Prigogine's dissipative structures as circular processes constantly turning new input into something familiar to keep the structure going. Without the constant input of the new (new energy, new thoughts), a structure will die. Too much new inflow would overwhelm and radically alter the system. The structure exists on the knife edge of this balance.

It is said that our brains make us more autonomous ("freer") than other creatures. For example, we aren't limited to living in one kind of society the way bees or ants or lions are; we can live in many patterns, even alone in caves. However, the brain which allows this remarkable autonomy is more open to fluctuation and therefore more unstable, unpredictable, than other elements of our selves (or other living creatures). Openness to fluctuation also means a greater, closer, more intimate connection with the flux and flow of the environment. So Jantsch found increased autonomy paradoxically related to an increased instability or openness, which widens and loosens pathways between what is "inside" and "outside" the structure. (Of course, on another level, there is no inside or outside to the structure anyway, since it's a constant flow of processes.)

To this paradox of autonomy can be added another. Physician and "biology watcher" Lewis Thomas has eloquently described in *The Lives of a Cell* how, like Alice's looking-glass cake, each of us exists autonomously only in the sense that we are passed around!

A good case can be made for our nonexistence as entities. We are not made up, as we had always supposed, of successively enriched packets of our own parts. We are shared, rented, occupied. At the interior of our cells, driving them, providing the oxidative energy that sends us out for the improvement of each shining day, are the mitochondria. . . . They turn out to be little separate creatures, the colonial posterity of . . . probably primitive bacteria that swam into ancestral precursors of our . . . cells and stayed there . . . with their own DNA and RNA quite different from ours. . . . My centrioles, basal bodies, and probably a good many other more obscure tiny beings at work inside my cells, each with its own special genome, are as foreign, and as essential, as aphids in anthills. My cells are no longer the pure line entities I was raised with; they are ecosystems more complex than Jamaica Bay.[44]

Jantsch's version of how in our autonomy we do not separately exist raises the stakes still further. It might be easier to grasp in a diagram.

Let's call the self-organizing (or dissipative or autopoietic) structure John Doe a square (no offense, John). Mr. Doe is autonomous. Despite what anybody says about him (such as calling him a square), he is his own reason for being. John, however, is composed of literally thousands of other autonomies of various shapes and sizes, such as those described by Thomas. There are also the heartbeat; electrical, brain, and muscle waves; and digestion. Each of these systems functions with relative independence. But that isn't all.

John is composed of systems which extend outside his physical body. He works for a corporation, lives in a city, and is part of a culture. Each of these entities, Prigogine and Jantsch say, is a dissipative autopoietic, self-organizing structure. It is born out of fluctuation and grows and maintains its dynamic

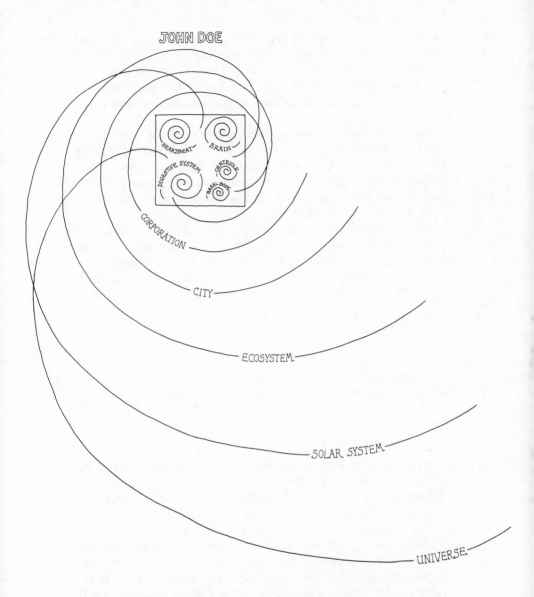

shape by a constant energy exchange and dissipation of entropy. Finally, of course, John is a part of a much larger
structure, the world's ecosystem, and beyond that, a part of
the solar system, galaxy, and universe—each of which is a
dissipative structure. In some of these dissipative structures
John is transforming energy flow into what is called John. In
others he is himself a flowing aspect of the energy exchange of

larger structures—his corporation, the city, his family, culture, religion. The diagram presents an extremely simplified picture. All of it, remember, is *in motion*. The galaxy, the city, the corporation, even the very cells of John's body are in a ceaseless energy exchange like water flowing through a vortex.

The order in this looking glass universe is not the same as the flowing interference patterns of Bohm's holographic mirror. However, on closer examination perhaps it's not really so different. Here too is a clear sense of flowing movement in which each dissipative structure is woven into others, observers woven into the observed.

Prigogine and his colleagues have carried us from the physics of thermodynamics through chemistry into the heart of biology. Next we'll consider the impressive potential impact of the dissipative-structure paradigm on what is easily the central theory in biology: evolution.

11. Looking-Glass Evolution

Most schoolchildren are familiar with the basic tenets of the theory of evolution proposed in 1858 by Charles Darwin and Alfred Russel Wallace and codified by Darwin the next year with the publication of his monumental *The Origin of Species*. Wallace and Darwin weren't the first to say that species continually evolve, but they proposed such a powerful explanation for how that evolution takes place that their thesis has remained cogent for over a century. The current version of the theory comes from a synthesis in the 1920s, '30s, and '40s of evolutionary theory with genetic theory and is known as neo-Darwinism.

According to neo-Darwinism, over the course of history new biological forms appear as a result of the struggle of individuals for survival. As individual creatures contend, those that are the fastest, fiercest, most deceptive, or otherwise "fittest" for obtaining nourishment and reproducing their kind will survive. "Natural selection," or in Alfred Lord Tennyson's famous phrase, "nature red in tooth and claw," will pick out the best. Although species may remain unchanged through thousands or millions of generations, processes at the molecular level of DNA ensure that a series of mutations is always present. Most mutations are ignored by the force of evolution because species are well adapted to their evolutionary niche. But when things begin to change—climate, vegetation, other animal species, and the like—these accidental variations may adapt more efficiently. Their genes will then spread through the entire population, because the entities having them will thrive and pass them on to the next generation. Previously thriving species may now find themselves maladapted and will die out. Continual changes in the environment make this process constant. Hence it is theorized that when the environment changed rapidly sixty million years ago, the dinosaurs, which had been admirably adapted, suddenly found themselves maladapted to the new conditions and so vanished in a very short time. The human species, it is said, has survived and flourished, virtually overrunning the planet, because its brain allows it to be infinitely adaptable to changing conditions and in some cases allows it to control them.

It is a simple and elegant theory: Chance variations selected for survival by a constant struggle for existence in a changing environment. What came to be known (unfairly to Wallace) as Darwinian evolution defeated the only major theoretical contender, Chevalier de Lamarck's theory that evolution takes place by passing on from one generation to the next what individual biological entities learn or "acquire" in their encounters with the environment. With the defeat of Lamarckism, Darwinism had no further "competition" for its own "survival" and quickly became one of the most successful theories in the history of science. As one writer celebrating the 125th anniversary of *The Origin of Species* wrote: "Darwin's theory is now supported by all the available relevant evidence, and its truth is not doubted by any serious biologist."[18]

Followers of Kuhn might note the unintended irony in this statement. Could anyone be considered a "serious biologist" who *did* doubt Darwin's theory? Certainly not creationists, who have picked up on the idea that after all, evolution is "not a certainty, but only a theory" to justify their own "theory" based on the Bible. Scientists naturally reject creationism as unscientific. Kuhn showed that while it may be that every scientific theory in the long run is only a perspective and not the absolute truth about nature, a successful theory must provide ample research puzzles. This creation theory cannot do.

Evolutionary theory has been successful precisely because it *has* supplied so many puzzles. In solving them, scientists have been able to establish intricate schemes of relationships among animals and plants and to make detailed speculations about the history of the planet from the first appearance of protein molecules through the development of complex animals. Seeds from this evolution theory have drifted over into other disciplines and taken root. Freud's idea of primitive instincts, for example, has a Darwinian origin. A serious rival to neo-Darwinism would obviously have to provide impressively good reasons for abandoning such a useful and pervasive paradigm. It isn't surprising that a serious rival has not appeared.

OBJECTIONS, HOWEVER

Nevertheless, not all scientists have been completely happy with the neo-Darwinian approach.

Some of the objections have been logical. The concept of "survival of the fittest" has been ridiculed as a tautology—the equivalent of saying that "the survivors survive." Observing a species that has survived, the evolutionary theorist says it must have survived because its shape was advantageous. Scientists then discover the advantage of that shape, thus adding the weight of new evidence to the conviction that species survive by adapting. The reasoning is circular and presumes what it sets out to prove.

Another problem appears, oddly enough, in what would seem the most obvious aspect of evolution—that life begins as simple forms and "progresses" over time toward those "more evolved," human beings currently holding the position at the top of this ladder. On the scale of individual species, the doctrine of survival of the fittest also seems to imply progress. Only the "best" among a given species survive, the cream of the crop. Darwin himself supported the sense of evolution as hierarchical. The theory's co-founder, Wallace, however, evidently imagined a nonhierarchical evolution, a process ceaselessly producing new varieties and species, no element of which is higher or more evolved than any other. Writer Arnold Brackman points out that in Wallace's travel notes he refers to the inhabitants of jungle and island cultures as "natives" while Darwin called them "savages." Victorian England had a difficult time accepting the idea that human beings evolved from animals and weren't products of a special creation by God. Darwin's version at least managed to preserve the conviction that man is the highest earthly form, and by implication the British Empire the highest civilization. That may have made the theory a little easier to swallow.

But there are problems with viewing evolution as an advancing hierarchy. For example, what if man, who has presumably achieved the apex of evolution, should fall off the ladder through his own violence, taking most of the lower rungs with him (except the cockroach)? Were we really more evolved? The evolutionary hierarchy seems to depend on our preference for thinking ourselves the most advanced of nature's creatures. This preference may be blocking a nonhierarchical appreciation of evolution. Granted, such an idea would be difficult to grasp, since we are so accustomed to ideas of power, superiority-inferiority, comparison. Everywhere one looks human beings have arranged the world into hierarchies and assumed that the greatest value lies at the top. The notion of a nonhierarchical order in which no level is more fundamental or higher than others is almost as unfamiliar as the idea of wholeness—*perhaps for the reason that wholeness itself is nonhierarchical.* The top doesn't really dominate the bottom, the bottom doesn't really build toward the top. We've have seen how Prigogine and Bohm both point to this conclusion.

The aforesaid difficulties with neo-Darwinism are logical and philosophical. There are also growing practical difficulties. Some of them center on the theory's reliance on DNA as the major mechanism for evolutionary stability and transformation. Most biologists believe that virtually everything about the shape and behavior of an organism will be explained by understanding the role of individual genes and the mutations of those genes. Therefore it has been rather unsettling for biologists to discover recently that in bacteria, DNA jumps on and off chromosomes, expanding and contracting so that, as the journal *New Scientist* says, "our very concept of a gene is now in doubt." As science writer Marilyn Ferguson put it, DNA may be more a "flux" than a code.[54] Nobel laureate and Vitamin C discoverer Albert Szent-Györgyi argues that the cell actually feeds information back to DNA and changes its instructions. Researchers have shown that some genes can convert into one another. If DNA is not a code but a flowing message, affected by the very things it is supposed to be controlling, then neo-Darwinians will have to face sticky questions about how natural selection works. How can environmental pressures select one gene over another if the genes themselves are in constant process?

A prominent biologist, the late C. H. Waddington, believed that genetic theory presents a picture that is far too mechanical and static to account for the amazing complexity and subtlety of life. His complaint with tying so much in evolution to DNA was that there are too few genes to account for the hundreds of traits exhibited by even the simplest organism. This is a particular problem when it comes to explaining how DNA could provide all the instructions necessary for the unbelievably complex process of developing an embryo, getting all the cells to divide and gravitate to the right places. (A looking-glass theory about this process is discussed in Part 4.)

Waddington believed the final form an individual embryo takes as it develops is not just given by a genetic blueprint. It is a result of the way genes interact with the environment. He suggested that the living system in its growth is like a river making its way downhill. If the river is dammed, it finds a new path for its flow. To illustrate this, he pictured what he called an "epigenetic landscape," a multidimensional world of hills and valleys.

The epigenetic landscape is a picture of both the individual organism and the external environment it develops in. Let's consider, for example, a developing fish. In the epigenetic landscape are what Waddington called "chreodes," well-worn pathways which represent the genetic tracks that a particular species of fish has followed in the past. When the fish egg and milt come together, the developing organism is set in motion like a ball rolling down the landscape. It will tend to follow beaten paths, the past genetic history. However, the landscape itself is in motion, like an ocean, shifting with predators and disease. The environment jostles the ballfish off its chreode and forces it to make a detour. To restore its internal

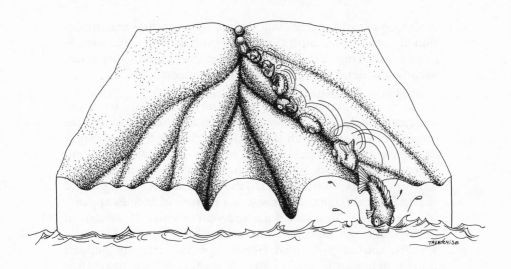

THE EPIGENETIC LANDSCAPE

energy balance, the ballfish pushes back and still reaches its final goal. The end result is an individual fish of the species. However, Waddington focused on the fact that this individual's development has subtly changed the landscape, subtly worn down a detour. If enough embryos are pressed by circumstances to make this detour in their own genetic structure, the detour may become a formal rerouting for the species. In this way, genetic structure and environment push against each other to create evolutionary types that are stable and always changing at the same time.

Harvard biologist Stephen Jay Gould and his collaborator paleontologist Niles Eldredge focus on other practical problems with neo-Darwinism. Gould in particular has become a powerful and articulate challenger of several important aspects of the current evolutionary paradigm.

Scientists have never seen an entirely new species created, though the evolution of many new varieties has been observed and new varieties have been created by animal and plant breeders. The most famous example of a naturally occurring variety witnessed by scientists is the peppered moth, which in less than a century developed a dark coloration that protected it when it lighted on tree trunks blackened by industrial soot. It has been assumed by neo-Darwinians that evolution of new species takes place by a gradual accumulation of such small adaptations until eventually a variety separates itself so much that it is no longer able to mate with other varieties which sprang from the same ancestor.

As Gould and Eldredge point out, however, the fossil evidence doesn't actually show a picture of gradual evolution. The geological record shows, instead, that when a species dies out it looks pretty much the same as when it first appeared. There are missing links between species. The inability to find these missing links is something of an embarrassment to paleontologists. Gould calls it "the trade secret." The picture the geological record *does* show is a picture of species appearing "suddenly" in a few thousand years (which is sudden in geological time), emerging into reality fully formed. Our own species, the large-brained *Homo sapiens*, virtually popped into existence amid several other hominids. Efforts to arrange the skeletal evidence of these hominids in a gradual sequence have not proved successful. There are too many gaps, and it is clear that some of the hominids were parallel evolutions, not ancestors of modern men. Gould and Eldredge call their theory of the sudden appearance of species "punctuational" evolution.

The two scientists also see other indications that evolution must take place in discontinuous jumps. Many organisms possess features that could not have been evolved one at a time in response to environmental stress. A bird requires a light bone structure and two wings. Wings in themselves wouldn't do; neither would a light bone structure. No single feature by itself would offer any survival advantage. For organisms to change their nature and function, many features have to evolve together. This means fantastic genetic coordination. Random mutation of a gene here and there couldn't accomplish such a transformation.

Moreover, Gould and Eldredge don't believe all the features of a species can be explained as adaptations. The two scientists provide another slant on the adaptation problem.

According to orthodox neo-Darwinism, if large changes occurred by an accumulation of small ones, then every difference between one species and another should be explainable as an adaptation. However, this isn't the case. Gould, who has an obsession for studying snails and crowds his office with their shells, offers a line of sea snails to illustrate this point. The oldest species in this line have simple sutures (joints) in the shells. Later species have very complicated patterns of sutures. Gould says all sorts of contradictory and unsatisfactory explanations have been given by evolutionists about why the complicated sutures are better-adapted and thus evolved. Gould believes the sutures may not have anything to do with adaptation at all but are aspects of the overall gestalt that happened when the new species made its jump to greater complexity.

Gould and Eldredge have been criticized for not adequately explaining how these sudden jumps from one species to the next take place. Critics charge "punctuated equilibrium," the formal name the two scientists give to their theory, with being too mysterious. The dissipative-structure paradigm may provide an answer to why and how these comparatively sudden and spontaneous jumps occur. However, from the point of view of the dissipative structurists, "equilibrium" in the theory title is an unfortunate word. In a dissipatively structured universe, evolution cannot take place and autopoietic structures cannot exist at all in equilibrium. But the argument here appears more semantic than real. Gould and Eldredge seem to use the word "equilibrium" to describe how species can remain unchanged for so long. So equilibrium in

Sea-snail sutures. Nautilus is the oldest species, Harpoceras the most recent.

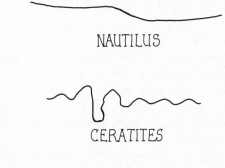

NAUTILUS

CERATITES

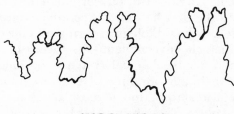

HARPOCERAS

this context seems to mean something like autopoiesis, the relative stability of a dissipative structure once it has formed.

Another proposal Gould and Eldredge make is to treat whole species as analogous to individuals within larger groups, such as families and phyla. Therefore, the whole of the species *Homo sapiens* would be considered an entity, as John Doe is an entity. Gould says the laws which guide the evolution of the *Homo sapiens* entity are different from the laws which affect John and Jane Doe. In other words, there are different levels of evolution, no level is fundamental, and one level does not obey the same laws as another, though the processes operating at each level (individual, species, family, phylum) feed back and forth in webs of relationships. Here, too, punctuated equilibrium looks like a Prigoginian approach. However, Gould and Eldredge retain the main features of Darwinian theory—the competition for survival.

Before we look at how one scientist has applied Prigogine's theory exhaustively to evolution and questioned even the primacy of the competition axiom, we should be aware of a caveat. As with quantum mechanics, the overwhelming majority of scientists are entirely happy with neo-Darwinism. Thousands of biologists successfully apply it to puzzles in their field. For these scientists there is no paradigm crisis. Far from it.

CO-EVOLUTION—OR WE'RE ALL IN THIS TOGETHER

Erich Jantsch began his professional career in his native Vienna as an astrophysicist, but from the outset his interests were wide-ranging. He was an accomplished musicologist, theater critic, engineer, businessman, student of English poetry, futurist, and consultant to numerous national governments on forecasting technology.

In the years before his death, he devoted his attentions to "systems theory," a powerful idea originated by Ludwig von Bertalanffy in the first half of this century. Von Bertalanffy was interested in how biological forms organize to sustain themselves in their environment. His theory developed the notion of open, nonmechanical systems continuously interacting with their surroundings. He also explored the interwoven relationships among hierarchies and how new complexity evolves. Though in itself his was not a looking-glass approach, it has had a major impact on looking-glass thinking. Erich Jantsch's last major project was a synthesis of Prigogine's theory with Bertalanffy's systems theory, along with physics and neurophysiology, urban planning, and a host of other disciplines.

Jantsch's monumental effort produces a gestalt shift—a looking-glass evolution. He called it "co-evolution."

Co-evolution is the dissipative-structure approach to the origins of species. Co-evolution considers neo-Darwinianism a limited picture of how biological forms change. It doesn't deny adaptation and the struggle of individuals for survival but rejects them as the overall driving force behind the development of new forms of life.

Scientist Gregory Bateson once described evolution as progressing from organisms like bacteria, which are "adjustors," able to match their bodies to the changing temperature of their surroundings, to "regulators," which developed complex mechanisms to maintain a constant body temperature, to "extra-regulators," such as humans, who need artificial environments to maintain a temperature range that is very narrow. The usual way of looking at evolution views each level of this progress as a move toward adaptation. But is it? The curious fact is, human beings are in some senses not as well adapted as bacteria. Jantsch asked pointedly: If the principle of evolution is adaptation, why did organisms grow steadily more complex? Consider the fact that your human complexity, which makes you extremely adaptable in one sense, also makes you as delicate as a piece of high technology. If you live in northern latitudes, think of all the trouble you had to go through to keep yourself warm last winter. A bacterium never shivers. Jantsch says of this, "The earliest life forms were by far the best adapted. *If the meaning of evolution was in adaptation and increasing the chances for survival, as is so often claimed, the development of more complex organisms would have been meaningless or even a mistake."* [Emphasis added.][23]

Darwinian evolution emphasizes adaptation to competition. In contrast, co-evolution emphasizes evolutionary cooperation—cooperation of a remarkable kind.

The basis of co-evolution is simple: *The development of structures in what is called microevolution mirrors the development of structures in macroevolution and vice versa. Microstructures and macrostructures evolve together as a whole.*

A way to grasp what is meant here by "micro" and "macro"

is to think of them as a ratio which can be applied to different scales. For example, a 1,000:1 ratio applied to insects might mean that a water beetle is 1,000 times larger than an aphid. One might be called a macro insect and the other a micro insect. If we take the cell as the macrostructure, then the molecules are the microstructure. If the macro is the species, then the micro is an individual within the species. If the micro is a phylum, then the macro could be the ecosystem.

Co-evolution says that changes which take place on the micro scale instantaneously effect changes on the macro scale and the reverse. Neither really "causes" the other in the usual sense. Microevolution doesn't build up in steps to create a macroevolution, nor do great shifts in macrostructures cause the micro world to respond. Each level is connected to the other by complex feedback mechanisms. They cause each other simultaneously. In effect there are no levels at all. They are all one dissipative structure—an evolutionary looking-glass cake. Jantsch said: "The one-sided application of the Darwinian principle of natural selection frequently leads to the image of 'blind' evolution, producing all kinds of nonsense and filtering out the sense by testing its products against the environment. As if this environment would not itself be subject to evolution."[23] Jantsch discovered a "two-sided" approach.

Erich Jantsch

We've already noted that for Prigogine and his collaborators the step between inanimate dissipative structures and animate ones (what the Chilean scientists called autopoietic structures) is a natural and inevitable one, not the "accident" or chance which orthodox biology assumes. As the looking-glass evolutionists see it, in the far-from-equilibrium conditions of the first years of earth—a bubbling caldron of gases and chemical reactions—dissipative structures abounded. In these, to borrow a phrase from Bohm, life was everywhere implicit. Very soon after the earth was formed and began to cool, chemical "matter structures" such as the Belousov-Zhabotinsky reaction appeared. There were autocatalytic chemical reactions able to reproduce themselves by copying with template molecules. They could pass on copying errors to the next generation of chemical structures. Such chemical "mutations" allowed the matter structures to engage in a kind of evolution.

Looking-glass theorizers think that rather than competing with each other for survival, these chemical matter structures actually evolved through a kind of cooperation. They flowed into one another, sharing the information of their chemistry. Cooperative exchanges led eventually to the formation of chemical structures containing nucleic acid (ultimately DNA) and the first appearance of "living" forms.

The early nonliving or not-quite-living microstructures cooperated rather than competed because they were part of a very large macro dissipative structure, the chemical system of the whole earth.

Lewis Thomas has remarked that earth viewed from space looks like a single cell and in fact functions very much like a unicellular organism. Similarly, the primordial earth was like a giant chemical reaction. This reaction regulated its activity by means of various autocatalytic chemical reactions which sprang up out of planetary fluctuations in the ocean and air. As these micro dissipative structures were subjected to further intense fluctuation they restructured themselves into greater and greater complexity, eventually evolving into what we would call life. As they evolved, they changed the chemistry of the macro (whole-earth) ecosystem so that it evolved. As this macro-level chemical ecosystem evolved, it in turn produced more and different fluctuations and new micro-level dissipative structures appeared. We can't ask which came first. Micro and macro produced each other, like reflections down a hall of mirrors.

This is the meaning of co-evolution. A structure doesn't appear in isolation either on the macro or the micro level but is a phenomenon born out of an environment in which everything affects everything, like Bohm's holomovement. Co-evolution is a description of a holistic unfolding, not an interaction of separate parts. In fact, the dissipative structurists use the word "unfolding" in a form which is actually quite Bohmian.

A little later in the history of earth, another unfolding of co-evolution occurred. At this point in time, ancestors of present-day bacteria swam in the seas and floated on freshwater lakes. Bacteria are single-celled organisms without a nucleus. They don't reproduce by transferring genetic material to the next generation the way cells such as our own do. Among bacteria there are no mothers and fathers to pass on their inheritance. There are only endless brothers and sisters, or rather "brosters" (or "sithers," if you prefer) since bacteria normally have no sex.

Instead of dying, the bacterial cell divides, making a DNA copy. As with a high-quality photocopying machine, there is no way to tell the duplicate from the original, though sometimes the process malfunctions and there are copying errors. In addition, there are also several different ways in which the bacteria can exchange genetic material, yet no new cell is produced by these operations. There are no bacteria species in a strict sense, since there are no breeding restrictions to separate different bacterial types. Though different strains of bacteria do develop in isolated situations, one bacterial strain can receive genetic information from another. The new booming genetic technology puts this capacity of bacteria to use in order to tailor new bacterial life forms. In effect, bacteria as a whole constitute a gigantic gene pool into which different types of bacteria may dip to get information they need to change themselves in changing situations. Mutations (copying errors) detrimental to survival are weeded out because the carriers of these errors die off before they can exchange them with many other of their sithers or brosters. Other "errors" are retained and become part of a genetic reserve fund distributed to many, many individuals, available for any contingency.

With the primitive bacteria a new level of co-evolution appears. Here the "macrosystem" is the totality of all bacteria and the "microsystem" is each individual bacterium.

Jantsch: "The evolution of this totality [macrosystem] only provides the possibility for the unfolding of the individual bacteria and the mutations occurring in this unfolding [microsystem] keep the overall system alive and dynamic."[23] In other words, the individual and the whole group exist by unfolding one another.

Somewhat later, co-evolution shifted to still another phase of complexity. Here, according to looking-glass scientist Jantsch, holistic evolution does something altogether implausible. It is as implausible as the nonlocality of quantum events. Nonlocality is a strange effect of holistic *space*. Jantsch unraveled a strange effect of holistic *time*.

So far as scientists know, only oxygen-breathing cells with a nucleus can form cell tissues and link with each other to create multicellular organisms. In the era of the bacterial dominance of earth, there was no free oxygen. Some of the bacteria responded to the fluctuations of their own genetic macrosystem (the gene pool) and to the fluctuations of the larger macrosystem of the earth by restructuring into forms capable of photosynthesis. For some 2,000 million years, photosynthetic bacteria performed the enormous task of totally transforming the atmosphere. According to Jantsch, there was a curious selflessness and vision in the way they went about it.

First, bacteria did not need to make this transformation in order to adapt to the oxygenless environment they lived in at the time. They were already well adapted to that environment (even today some strains can survive only in oxygenless places such as mud or our intestinal tracts). The primary advantage seems to have been that the presence of free oxygen makes bacteria fifteen times more efficient in metabolizing glucose. However, that efficiency can only be achieved *after* the oxygen is freely available. That's a problem. How could did the bacteria "know" they would be more efficient if they all worked together to produce enough free oxygen so that large numbers of them could later take advantage of it (by changing one step in their metabolism)? This is only the beginning of the mystery.

Bacteria using free oxygen are most efficient at a 10 percent concentration of oxygen. If adaptation is the criterion, why did the early photosynthetic bacteria go on to produce an atmosphere containing twice that amount of free oxygen? This meant *decreasing* their own efficiency. Yet engaging in this maladaptive activity created just the conditions necessary for the development of oxygen-breathing organisms with cell nuclei and sexual reproduction. The atmosphere the bacteria created was thus fortuitous, not for bacteria but for the whole earth and its future evolution. This suggests bacteria were not acting out of a process which optimized their survival, but out of a deeper process—one that could leap across time to "anticipate" what was required for the further unfolding of evolution.

On a grand scale, bacteria in this process were performing a function similar to the autocatalytic step we observed in the Belousov-Zhabotinsky reaction. At a primordial moment of evolution, they entered into a cycle around which the dissipative structure of the earth's ecosystem itself began to turn, and they continue to act as catalysts and stabilizers for the whole flowing structure of the planet. A new theory describes how bacteria do this.

The conventional view of the earth's ecosystem is a static or equilibrium one: a concentration of about 21 percent oxygen, 79 percent nitrogen, and trace gases. A theory called the Gaia hypothesis (after the earth mother in Greek mythology), proposed by American microbiologist Lynn Margulis and British chemist James Lovelock, argues that the equilibrium view is inaccurate. They point out that none of the atmospheric gases is ever allowed to settle into equilibrium because living organisms are continually engaged in breaking the gases down and re-creating them. There is a constant flow of the atmosphere through life and back into atmosphere so that some gases (such as ammonia) exist in much higher concentrations than they would in a system whose particles are slowly randomizing and drifting toward equilibrium.

The Gaia (earth) system, including life forms, has been compared to a bathtub with the same amount of water running into it as is draining out from the bottom. But the image is too mechanical. The Gaia system is a giant autopoietic or self-organizing structure like a living organism, containing far-from-equilibrium states which foster the emergence of new micro self-organizing structures. That's the view looking down—from the macro level to micro levels.

We can also look from the other direction: The primitive photosynthetic bacteria played a crucial role in creating the Gaia system, and bacteria continue to play that role. For example, lichens are single-celled bacterial algae linked with fungi in symbiosis. The algae and fungi, which in other circumstances would exist separately, combine to form a partnership. The algae supply the photosynthesis and the fungi contribute water, carbon dioxide, and a firm grip. This partnership turns the equilibrium structure of rocks into earth and minerals, which in turn enter into the nonequilibrium cycle of life, become life, as plants. Bacteria are also present in virtually every cell of complex organisms such as ourselves. They exist there in the form of centrioles and basal bodies, autonomous entities which aid in a constant flow of gases and matter through the Gaia system.

So the cycle moves from micro to macro and back. Micro bacterial self-organizing structures turn the planet into a huge, living macro self-organizing structure which, in turn, creates fluctuations that encourage the appearance and maintenance of self-organizing structures on the micro level. Small wonder co-evolutionists think the Darwinian concept of organisms struggling to adapt to their environment is one-sided.

If the law of evolution is not the adaptation of individuals for survival, what is it? Jantsch believed that the primary law and purpose of evolution is openness, an expanding holistic multidimensional web of processes built in all directions by the macro-micro co-evolution of systems. The goal of this web is what Jantsch called "the extraordinary intensification of life."

In looking-glass evolution the death of species or ecosystems isn't viewed as entities failing to adapt. Species appear and vanish as aspects of the general co-evolutionary unfolding. The death of the whole ecosystem of jungle vegetation which supported the dinosaurs was a crucial aspect of the unfolding of mammals and eventually man. Moreover, for the past hundred years, the residue of this lost ecosystem (coal and oil) has made possible a dramatic new evolutionary development—the industrial revolution. Throughout evolutionary process, earlier moments seem to presuppose moments that will come after, taking directions that will facilitate further growth. So while some species may die out, evolution *as a whole* will be expanded. From the co-evolutionary point of view, past and present seem to exist together in a higher-dimensional reality we call the future. (Does this suggest Bohm's higher-dimensional fishtank?)

However, Jantsch wasn't suggesting that a co-evolutionary

universe is a universe unfolding according to some foreordained, deterministic, or God-given plan. He and Prigogine compared their theory to the Greek idea of the world as a work of art, and contrasted it to the usual scientific idea of the world as an automaton. A work of art is a creative order. For Prigogine, what happens at the bifurcation point where dissipative structures are formed is the creative moment, a macroscopic "uncertainty principle" equivalent to Heisenberg's microscopic uncertainty principle. The observer must accept he is no longer dealing with a mechanical order that can be totally determined. He inhabits an indeterminate whole which exists beyond formulation of any particular level. In this way, the universe is as free from ultimate interpretation as a Bach cantata or a poem by Blake.

Obviously, a wide gulf separates the cooperative evolutionists of the dissipative-structure paradigm from the mechanistic view taken by most contemporary biologists. As mentioned, orthodox molecular biologists believe the shape life forms take is controlled by the information in the genes. For the co-evolutionist, however, form is not given or even directed by genes. The existence and form of an organism (that is, the organism as a dissipative co-evolutionary process) develops *using* genetic information.

The question of whether the mind is something more than a brainful of electrochemical impulses or exists beyond the brain is resolved by a similar argument: Mind is a process structure woven inextricably into other process structures and includes, but is definitely not limited to, the chemistry in our skulls.

When bacteria began to produce free oxygen, they created a bifurcation point of intense fluctuation that led to oxygen-breathing forms of life and a new level of complexity for the ecosystem. According to Jantsch and Prigogine, the human mind, which was itself created by the co-evolutionary fluctuation of the Gaia system, has brought on a new planetary bifurcation point. Human brains are producing huge fluctuations by restructuring the earth's environment. They are even beginning to intervene directly in evolution by technologically creating new kinds of life.

Does this mean human beings with their complex minds have climbed to a higher level than bacteria? For Jantsch, the Darwinian hierarchy is a narrow vision based on belief in a universe of separate parts. Cooperative evolution portrays man as being neither higher or lower. As with our fictitious John Doe, we are process structures which exist as many levels of microevolution (including bacteria) and also as levels of macroevolution (including the Gaia system). We can be called higher only in the sense that we are more autonomous as individuals. Jantsch saw greater individual autonomy as the ultimate direction of co-evolution—presumably bearing in mind the paradox that for a self-organizing structure, the greater one's autonomy, the more extensively one is being passed around.

It should be evident that the idea of cooperative evolution offers an immense, paradigm-shifting explanation for the discontinous jumps in evolution—why species remain unchanged for long periods and then suddenly diverge into new species, why and how the sutures of sea snails grow more complex. Co-evolution overthrows neo-Darwinism and asserts that life forms are not created piece by piece in small changes: They're dissipative structures arising spontaneously and holistically out of the flux and flow of macro and micro processes. Co-evolution explains the gentleness of the whale, the delicacy of the tropical fish, the gay markings of the butterfly, and the curiosity of the human mind not as simply responses to the demands for survival but as the creative play and cooperative necessity of an entire evolving universe.

THE LIVING UNIVERSE

In his labyrinthine *The Self-Organizing Universe*, Jantsch traced co-evolution from the first years of the planet to the present, showing an ever-expanding complexity of forms interpenetrating one another and creating the whole earth, in Thomas's phrase, as "a single cell." He also applies co-evolution theory to the rise of galaxies and star systems and the birth of the universe itself. Finally, picturing the universe as a cosmic dissipative structure, he brings the dissipative-structure paradigm full circle to challenge the basic conclusion of classical thermodynamics—that the universe is running down.

According to Clausius and Boltzmann, overall entropy cannot be reduced. Dissipative structures, remember, are so named because they must dissipate huge amounts of entropy

Does nature recycle entropy into new order?

to stay alive. For every gain in structure, the surrounding environment must pay with a corresponding increase in chaos. But perhaps not.

Lichens attack the equilibrium structures of rocks and convert them into far-from-equilibrium situations which will give rise to new order. What is entropy and waste to one system may become nourishment to another.

Jantsch, the astrophysicist, proposed that recycling ultimately overcomes the move toward equilibrium because it takes place on a cosmic scale.

Physicists know that the universe is expanding from the "big bang" explosion of primitive nuclear forces. They believe it will either continue to expand forever or that at some point the force of gravity will begin to pull it back together again, like the slices of a looking-glass cake.*

---

* Discovery in 1983 of huge gas clouds floating between galaxies, the so-called "missing mass" of the universe, supports the hypothesis that the force of gravity will eventually prove strong enough to cause a cosmic contraction.

The latter possibility would mean the universe is not running down. It is expanding and contracting like a heartbeat. Jantsch speculated that as it beats back, it will generate a new fluctuation, a new micro universe, a new far-from-equilibrium explosion, giving rise to new macro dissipative structures like galaxies, planets, cells. In Jantsch's scenario, the universe as a whole is the ultimate of all dissipative structures—*feeding off the far-from-equilibrium environment of itself.* "Life appears no longer a phenomenon unfolding in the universe—the universe itself becomes increasingly alive."[23]

From the cosmic scale of the expanding and contracting dissipative universe, we complete the cycle—returning finally to the smallest level. Contemplating this level, one group of physicists has invented a theory of subatomic interaction which mirrors the cosmic scale of co-evolution. Developed by Geoffrey Chew in the early 1960s, Bootstrap theory, as it is called, takes its cue from those looking-glass features of particles and quarks splitting up and dividing back into themselves. The name of the theory comes from the eighteenth-century teller of tall tales Baron von Munchhausen, who was reputedly able to lift himself up by his own bootstraps. In an analogous way, particles are said to be able to give birth to themselves. (See the illustration on page 77.)

Bootstrappers claim that each elementary particle *consists* of all other particles. No particle is really more fundamental or elementary than any other particle. In fact, Bootstrappers do not think of particles as separate entities at all. What other physicists record as particles Bootstrappers treat as intermediate states in ongoing process webs of energy. Instead of trying to develop a single mathematical approach, as the grand unificationists are doing, Bootstrap theorists are attempting a number of different overlapping mathematical models to create a web of relationships, none more fundamental than any other.

Thus what is fundamental in the universe of Prigogine, Jantsch, and the Bootstrappers is not individuals, species, phyla, galaxies, or elementary particles, it is the whole. For them, from micro to macro and back, a spontaneous, flowing whole of process reigns.

12. Some Implicate and Dissipative Questions

Ilya Prigogine's insight into the sudden appearance of collective order in nonequilibrium thermodynamics has spread into many areas. The mathematics of dissipative structures has been applied to urban development and the fluctuations of financial markets. The U.S. Department of Transportation uses it to predict traffic-flow patterns. Sociologists employ the theory to describe social changes. Psychologists are considering the sudden emergence of altered states of consciousness as dissipative formations. Medical scientists are studying the nature of the fluctuations that cause the "normal" micro cancer cells in the body to spontaneously develop into macro cancer. And Prigogine himself shuttles between his centers at the Free University of Brussels and the University of Texas at Austin, coordinating research.

The impact of the co-evolution side of the paradigm has, as yet, not been felt, but its potential for altering our view of reality is enormous. Darwinian theory had a profound effect on the perception of man's "animal" nature. In the late nineteenth century, "social Darwinism" was used to justify cutthroat business practices, and even today a "survival of the fittest" attitude prevails in many areas of society. At this point, one can only wonder about the possible effect of a theory which emphasizes the cooperative aspects of evolution and depicts us as intimate participants in the fate of all nature.*

CONTROVERSIES OVER CONCLUSIONS AND TERMS

Ironically, critics have charged the Prigogine paradigm with precisely the opposite thrust. The paradigm argues that fluc-

---

* We should note, however, that there has recently been a spate of new theories which look suspiciously Jantschian. One, for example, argues that the ecology as a whole uses human beings to sustain itself, somewhat in the manner an intestine uses bacteria for digestion.

tuation leads to higher levels of complexity. Does that imply that the technological fluctuations which precipitated the miseries of the industrial revolution, economic upheavals, social conflict, even war can be justified as inevitable and ultimately positive because they further evolution? Is the wholesale technological manipulation of nature an unstoppable aspect of our role in the new evolution of the planet? Does the fact that a relatively small fluctuation can become the dominant factor in creating a new system imply that a creative minority of humans is about to transform society to a higher plane?

Jantsch, Prigogine, and other followers of the dissipative-structure theory have drawn such conclusions, though the fallacies seem obvious. The creative minority which transforms society might be Plato's philosophers—or it might be the Nazi Party. Technology, as Jantsch himself pointed out, often intrudes on co-evolution by leaching novelty out of natural systems and driving them into equilibrium. (An example would be the overcultivation of cropland which turns it into desert.)

We note that Bohm's approach differs substantially from that of the dissipative structurists when it comes to such questions. From his point of view, for example, genetic engineering may or may not have benefits, but humans will not be able to judge as long as they treat the universe as parts. But what would a technology be without parts? A holistic technology? The very idea transforms the discussion.

Bohm's acute awareness of the pitfalls of language and the subtleties of wholeness derives from his long-standing belief that metaphysical and physical issues cannot be separated. In contrast, Prigogine's paradigm emerged almost exclusively out of attempts to solve particular scientific problems and arrived at metaphysical issues secondly, as a consequence of its discoveries.

Critics of dissipative-structure theory have also complained that while Prigogine's mathematics may be unassailable, some terms in the paradigm (particularly Jantsch's version of it) are vague. The terms "higher" and "lower," "complexity" and "levels," carry a heavy baggage of everyday connotations that can easily obscure the context in which the theory uses them. There, they can only be very relative terms. Complexity, for example, is not necessarily a "higher" level of evolution. In technology the first models of devices are often complex and cumbersome. More advanced (or evolved) versions are simpler, more elegant. Complexity, like the concepts of adaptation and hierarchy, is in the eye of the beholder.

An even greater ambiguity surrounds the theory's crucial term—"fluctuation."

It appears at first glance that Prigogine's fluctuation—that environment which breeds dissipative structures—is the conventional idea of pure randomness, primal chaos, order appearing spontaneously and inexplicably out of disorder. It is an idea which, as Ferguson says, "seems outrageous, like shaking up a box of random words and pouring out a sentence."[54]

Fluctuation, however, has the root "flux," flowing. Bohm's paradigm views flux not as randomness but as higher-dimensional order, hidden order. Jantsch asked: "What is the meaning of 'randomness' in the context of a multilevel evolution in which each level brings new ordering principles into play? How random is the fluctuation which is introduced into a system by one of its members or by an outsider if this individual is itself the product of a long evolutionary chain?"[23] Fluctuation on one level is order on another. On this, it appears, Bohm and Jantsch agree.

Jantsch also declared: "It seems that we frequently confuse indeterminacy and chance. Indeterminacy is the freedom available at each level, which, however, cannot jump over the shadow of its own history. Evolution is the history of an unfolding complexity, not the history of random processes. . . . Nothing is random but much is indetermined and free within limits."[23] How can something be both free and limited? Jantsch's answer seems to be a variation of Bohm's. Since all processes are linked together, there is no bottom or top; it is the whole which determines what happens and the individual in its freedom is an aspect of the whole.

Another indication that the dissipative structurists' fluctuation is not just the chance gurgling of order out of disorder lies in a strange phenomenon: If a dissipative structure is forced to retreat in its unfolding—for instance, if there is a significant reduction in the energy it needs to maintain itself—the structure doesn't just fall apart chaotically. It follows a path back to earlier stages of its evolution, almost as if it "remembers" them. Read the steps in the Belousov-Zhabotinsky reaction backward to see an example. This seems clearly to imply the folding and unfolding of layers of hidden order. Again, an echo of Bohm. Perhaps the uncertainty in the theory about whether fluctuation means randomness or hidden order is a clue; are scientists on the verge of a new understanding of randomness, some new way of perceiving how fluctuation is both randomness *and* order? At present, it's unclear.

The idea of a "process structure" also has ambiguous dimensions. In the dissipative-structure paradigm, process replaces the conventional idea of "things." Process gives priority to time rather than space, which is appropriate to a paradigm so concerned with time. But process may still imply things and parts, not spatial but temporal parts. Dissipative structurists don't always appreciate how treacherous the ground is surrounding any discussion of parts vs. whole.

JOINING PARADIGMS

In a compelling number of ways, the dissipative-structure/co-evolution paradigm seems parallel or actually convergent with Bohm's implicate order. One can only speculate whether these two looking-glass theories can in fact be combined. But the speculation here is tempting.

○ Is the sudden appearance of a dissipative structure out of far-from-equilibrium fluctuation actually a picture of an implicate form unfolding from one dimension to another?

○ When a dissipative structure undergoes intense fluctuation, it spontaneously evolves into a new level of complexity. Could this be an indication that an unknown order exists (as Bohm claims) in different dimensions and that these dimensions can unfold one after the other into our familiar three dimensions? When a "higher-dimensional reality" unfolds into our three dimensions, does it appear to us as a new order having "greater complexity"? That is, is "greater complexity" really the three-dimensional way we perceive higher-dimensional order?

○ Is the co-evolution of forms a law of the implicate order—a way of describing how different implicate ensembles affect each other so that they necessarily unfold together?

○ Prigogine says no level of description or law is fundamental. We move from one level to another, but no overall law (what Bohm calls the law of holonomy) covers all levels. This means that between different levels of description there are gaps, such as the gap between the reversible laws covering the Schrödinger wave equation and the irreversible laws governing cats and human observers. Do these "gaps" indicate "places" where the explicate order of one level is in some way winding down into the implicate order and then reappearing as an explicate order at another level? If so, this would suggest the various levels of explicate order (such as particle level and cat level) are literally divided and connected by the implicate order. (And, of course, it would also suggest how implicate and explicate are inextricably wound into each other.)

○ Bohm's vision of semiautonomous subtotals seems related to the ideas of autonomy advanced by the Chilean biologists and Jantsch in the theory of autopoiesis. Is autonomy of subtotals (like ourselves) a paradoxical destiny and inevitable aspect of the movement of the whole?

○ Are Prigogine's "bifurcation points," when systems can go in any direction, points where different implicate ensembles converge? Put another way, are the bifurcation points really points where an underlying unity exists among several possible forms, only one of which will actually unfold and appear into our three-dimensional (explicate) reality.

○ The flow-through order of a dissipative structure which appears out of chaos or fluctuation connects it to all its surroundings, eventually (according to co-evolution) even to the appearance of other dissipative structures on other macro and micro levels. Can this idea be united with Bohm's holographic idea of everything enfolding everything else?

Such speculation about points of contact seems natural and one feels certain these two looking-glass paradigms can be connected in other ways as well. Bohm's colleague Basil Hiley has recently indicated an attempt to correlate or transform the two theories, perhaps uniting Grassmann algebras and the René Thom catastrophe theory. Prigogine's science of becoming and Bohm's and Hiley's interest in the algebras of becoming suggest they might possess solid common ground.

Appearances are one thing, however, and the realities of science something else. The two theories may eventually come together or further work may reveal that they are in a deeper way irreconcilable.

For the present, we will have to be satisfied with savoring what we can of this rich new taste of looking-glass cake. At the very least, we perceive it contains familiar ingredients. The theories of dissipative structures and co-evolution portray, as did the implicate order, a holistic universe where everything affects everything else—an alive, multidimensional, creative reality where the observer is the observed, the laws of nature evolve, and wholeness is a flowing. Thanks to Ilya Prigogine and his colleagues, we have been granted a new picture of the sudden vortexes described by Bohm and have traveled by a strange new path to the timeless river from which they spring. But we are not done yet finding ways in which the wholeness of the looking-glass may be strange.

4

Rupert Sheldrake
Seeks Hidden Forms

This sounded nonsense, but Alice
very obediently got up, and carried
the dish round. . . .

13. Why Is This Cell Here, Not There?

Our third looking-glass expedition will be comparatively brief. It has only just begun. It's not yet clear if it will find what it's looking for. What it is looking for is clear: proof that there are hidden forms—forms which, while they are themselves beyond space and time, give a shape to our space-time world of things.

The current leader of this expedition, which commences in biology and extends into physics, is a young British biologist, Rupert Sheldrake. His expedition may, in the future, produce spectacular results. It may provide a concrete insight into how a dissipative or autopoietic structure, once it emerges, remains stable over time, across generations. It may provide a picture of some of the laws by which an implicate ensemble becomes explicate and then implicate again. It may provide, that is, another perspective on the looking-glass. Or it may end up as merely a curious footnote in the annals of scientific theories—a wild, speculation which failed to find factual support.

THE PROBLEMS OF FORMING

The problems Rupert Sheldrake tries to answer with his bold and curious theory are problems of "morphogenesis." The word comes from Greek roots: *morphé* meaning "form" and *genesis* meaning "coming into being"—the coming into being of form, the process by which things attain, maintain, and pass on their forms. Sheldrake has confronted six main problems of forming.

The first is a very old one and can be put as a question, one a child might ask: "How does that tiny acorn become a mighty oak?" If you think about it, this is really rather a remarkable question. How does anything grow? What is it that directs the transformation which takes place as embryonic structures turn into mature ones? When things grow, they don't just get bigger. The human being in its first days is a ball of cells which somehow metamorphoses into an intricate structure of limbs, lungs, brain, skin—all the myriad different types of cells which make up a baby's body. This body continues to change even after birth. How did one cell in that first ball of cells know to divide the way it did, to shape itself in a particular way and gravitate to a particular place amid the bubbling cacophony of other dividing cells? Why is that cell here on the tip of the nose, not somewhere else? And how did the cells somewhere else know to be where *they* are? As a zoologist once remarked, "The growth and development of any living system would appear to be controlled by someone sitting on the organism and directing its whole living process."

The second problem involves what is called "regulation." If part of an embryo is removed or a part is added, the organism continues to develop a more or less normal structure. If one of the first two cells of the sea urchin embryo is killed, the other doesn't produce half a sea urchin, it produces a *whole* but

The drawing illustrates regulation. On the left is the embryo of a normal dragonfly. On the right, the top half of a dragonfly egg has been tied off. The embryo has regulated its forming process to this interference so that in the bottom half a small but complete embryo is formed.

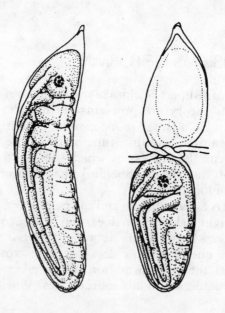

smaller sea urchin. Conversely, if the first cells of two sea urchin embryos are put together they will unfold into a single giant sea urchin, not Siamese twins. As cell regions are differentiated, the embryo will lose this capacity but the existence of regulation at early stages suggests that something is directing the process. It appears that something must be guiding the organism to its morphological goal—a single whole entity of a certain shape—even when interfering scientists put up detours.

The third problem is regeneration. In high school biology class you might have seen the following demonstration: A flatworm is sliced into pieces and the flatworm doesn't die. Instead each piece magically regenerates into a whole worm. It is well known that some nerves in the human body will regenerate, as will your skin when you are cut. In one experiment a scientist surgically removed the lens from a newt's eye in a way that could never occur by accident in nature. He did this to rule out the possibility that regeneration could take place according to genetic instructions which had been selected for as an adaptation to the demands of the environment. This is the usual explanation for regenerative capacity and is supposedly the explanation for why your skin regenerates: Natural selection has favored individuals who could heal wounds, and genes were selected which could instruct this process. However, the newt's case poses a problem. In a normal newt embryo, the lens develops by unfolding from the skin. Once the newt is mature, this path of lens formation is blocked. When the scientist cut the lens out of a mature newt, the lens nevertheless regenerated by developing from the edge of the iris. Somehow the organism possessed a forming process that could fill the hole. How is this possible?

The fourth problem of morphogenesis is reproduction. How do two parts—sperm and egg in humans—manage to become a whole, with a shape completely different from that of the parts?

The fifth problem extends the idea of forming beyond the processes which *shape* matter and into those which also *move* it. Sheldrake asks: How is it that living things move the way they do? Plants and animals have characteristic types of motion. Some tropical plants, for example, move to slant their leaves away from the burning equatorial sun. Shaded plants do the opposite and orient to catch the maximum number of light rays. Animals of all sorts have typical feeding patterns.

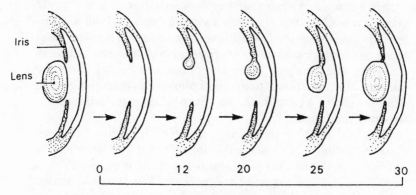

Days after surgical removal of lens

After the original lens was surgically removed from the newt's eye, a lens regenerated from the margin of the iris.

There are typical movements for each species associated with the acts of reproduction and growth. An important insight of Sheldrake's hypothesis is his connecting the processes which form matter and those which propel it habitually through the environment.

The sixth problem is another schoolchild's question: How did the giraffe develop its long neck? Was it from straining to eat leaves on the tops of trees? How did the camel get the calluses on its knees, which are present even in its embryo?

These last are, of course, the very questions Darwin and Wallace attempted to answer with the theory of evolution. In fact, neo-Darwinists believe they have the answer to all the questions Sheldrake poses.

For the neo-Darwinian biologist, the growth of the embryo is like the growth of a car on a computer-controlled assembly line; each part fits exactly into its neighbors as determined by the computer program. On the biological assembly line, that program is DNA, the blueprint which administrates such factors as shape and instinct, reproduction and regulation and regeneration. It is the ultimate determiner of the forming process at all levels. Environment may have an effect upon the final size of an organism or produce defects in its metabolism, but no matter how many generations are affected in this way the genetic material in each cell does not change. New forms are possible only through the random mutation or exchange of amino-acid sequences on the DNA chain. No matter how many generations of camels kneel in the sand, an acquired callus will never be passed on. Acquired calluses can have no effect whatsoever on the genetic structure of the cells. Secure in this paradigm, the conventional biologist would be confident in asserting that all the processes Sheldrake describes as problems, indeed every aspect of a living being, can be reduced to a set of physiochemical reactions and nothing more.

Young Rupert Sheldrake received this message as a student at Cambridge University in the early 1960s and wondered about it. Clearly biological reductionism explained a great deal about the world of living entities. But, he asked, is it really such an exhaustive description?

While at Cambridge, Sheldrake joined an informal group of scientists and philosophers who met to discuss and debate various questions. As a member of this group, Sheldrake was cast into the role of defending orthodox biology. The effort to hold up neo-Darwinism against astute minds strengthened his suspicions about its weaknesses.

One of the major weaknesses, he realized, is the claim that DNA controls forming. Biologists might find or hope to find the ultimate atom (or gene) of many traits in living organisms, but would they ever find the atom of forming? To Sheldrake it seemed unlikely. He reasoned that since every cell in the organism has exactly the same DNA code, it is difficult to see how this identical DNA configuration in each cell could tell one cell to be a brain cell and another to be a muscle cell and direct each of them into one particular site and not some other.

A second weakness is the reductionist approach to form: As it stands now, the thousands of coupled chemical reactions that power a single cell lie far beyond any mathematical solution with even the largest computer. Yet molecular biologists maintain complete confidence such solutions are possible. Sheldrake realized that even the limited solutions which biologists have come up with so far involve numerous arbitrary factors and are not so much predictions as descriptions after the fact. For example, the laws of thermodynamics dictate that a molecule will assume a shape that requires the minimum amount of energy to maintain. When molecular biologists develop formulae to calculate what this shape should be for a given molecule, they first already know what shape it is and they feed this information into the calculation. This is something like a weather forecaster being able to describe perfectly an airflow pattern and the weather it produced yesterday. When it comes to actual predictions of what forms things will take, there often seem to be hidden variables at work.

A third weakness Sheldrake perceived involved that by now familiar question—adaptation. Neo-Darwinians claim the camel's knee callus is an adaptation resulting from the selection of some random mutation. But while a knee callus may be a minor convenience, it is difficult to see how it could be a major advantage in the fight for survival.

Sheldrake saw that biologists were taking it as an article of faith that the final form of an organism and its functioning could be explained strictly in terms of genetic, physiochemical reactions. He felt the real evidence that this faith would be justified is thinner than its believers suppose. Sheldrake concluded that the admitted successes of reductionist biological theory were obscuring the fact that large areas of ignorance were being papered over by dogma and hope. As far as he was concerned, his six problems of morphogenesis were not being answered by mechanical DNA theory and there was no good possibility that they would be. In physical equations, all the terms can be accounted for: mass, energy, momentum. Not so with form. Take a simple example: If you burn a flower to ashes, the mass and energy are conserved and can be reflected in the scientists' calculations. But form is not a conserved quality. It is simply destroyed. Where does it go? Sheldrake perceived that while there might be no crisis for biologists generally in their paradigm, there was a crisis for embryologists. Embryologists, as scientists who deal directly with questions of forming, are, in fact, at an impasse. They have no good theory to explain what they see taking place in even the simplest organisms—the miracle of development and growth.

Waddington's idea of the epigenetic landscape seemed something of a help because it was less mechanistic. Yet, in another sense, it was largely a recasting of earlier notions and didn't go far enough. Sheldrake decided Waddington's model did, though, provide a good way to focus the problem. For example, the ball (developing organism) rolling down the landscape in its chreode is being attracted by the future or end point. This end point is the developed form. How does this attraction of the present toward the future take place?

In the mid-1960s, the young biologist took a year away from his research to read philosophy at Harvard. During this period he studied Goethe's writings on science and the French philosopher Henri Bergson on vitalism, and he entered the mazes of Alfred North Whitehead's holistic philosophy in *Process and Reality*. He was particularly drawn to the vitalist position. The term "vitalism" is associated with Bergson and a school of philosophers in the 1920s. Bergson argued that a life force or *élan vital* exists in biological entities which can't be quantified. Mechanism, such as the biological reductionism we've been discussing, holds that ultimately everything can be quantified. Bergson insisted that quality should be recognized as irreducible to quantity. The debate between mechanist and vitalist approaches is an old one. On the vitalist side have been ranked Pascal, Goethe, and Bergson; for the mechanists, Descartes, Newton, and Darwin.

Centuries ago, what we now call science was a branch of philosophy. And at times of crisis it seems natural for scientists to return to these roots. We have seen how Heisenberg and Einstein found a grounding in philosphical ideas that guided them on their voyage to contact nature. But the message which the activity of philosophy itself conveys is more important than any particular philosophical ideas a scientist might embrace. Despite the wide differences in their philosophical conclusions, all great philosophers have shared a willingness to set aside prior assumptions and look at problems freshly. This willingness to let go past assumptions has been the starting point for every major philosophical idea. So, although Sheldrake has said his reimmersion in philosophical thinking convinced him there is a sharp line to be drawn between physical and metaphysical ideas in science, his study of philosophy also prepared him to do what both metaphysicians and scientific theoreticians do best— think through the problem from the beginning.

After a period of research into cell biology as a Research Fellow of the Royal Society, in 1974 the young scientist moved to India, where he investigated the physiology of crops. By now his ideas on the formation of living systems had crystallized, and toward the end of the decade he retreated for several months to a Christian ashram in southern India where he wrote *A New Science of Life*.

MEASURING LIFE'S FIELDS

In the hypothesis he developed, Sheldrake proposes that there exists a state which mediates between DNA and the forming processes of an organism. This mediator is a complex set of hidden fields which direct all stages of morphogensis and the final forms things take, including their behavior. He calls his insight into these hidden variables of forming "the hypothesis of formative causation."

The idea of fields is itself not an original one. Bergson had proposed the *élan vital*. Biologist Hans Driesch had argued that the fate of any group of cells in an embryo is determined not only by the genes but by a guiding principle outside the cells he called an "entelechy." More recently, claims have been made of so-called "auric" fields seen to exist around living forms by people with psychic powers. In the 1960s and '70s, the existence of these fields appeared to achieve an uncertain scientific legitimacy with the discovery of Kirlian photography, which shows living things emanating flamelike radiations. Kirlian photographs seem to reveal that these fields shrink and evaporate if, for example, a leaf is cut and dies. Later, some scientists countered that Kirlian fields were only effects resulting from the experimental apparatus and photographic methods used.

A more scientifically rigorous study of fields in living organisms was conducted by a Yale biologist, Harold Saxton Burr. Beginning in 1935, and for almost forty years until his death, Burr explored what he called "L-fields." Using a special voltage detector which does not drain away electrical output with its electrodes and so is considerably more sensitive and less intrusive than the usual means of voltage detection, Burr discovered that different types of organisms— trees, slime molds, human beings—have identifiably distinct patterns of electrical activity. His instrument also showed that individuals possess characteristic fields, like electrical fingerprints. Disruptions in this characteristic pattern, he claimed, foretell events that will show up in the physical structure of the organism. For example, a field change may indicate the growth of a cancer. (The same claim is made for Kirlian fields.) In human subjects, Burr discovered, changes in the fields registered changes in psychological mood as well as health. In women he studied, Burr found that the exact moment of ovulation could be predicted by a change in the field voltage. His researches concluded that many of the women he studied did not ovulate in the middle of their menstrual cycles, and the voltage change was successfully used to advise women who were thought to be infertile about when they might conceive.

Burr also discovered that there are fields and field changes associated with an organism's movement in the environment. He speculated that L-fields governing different parts of an organism or different stages in the development of embryos are connected to each other in a hierarchy and that these fields are in turn affected by other fields in the environment, including such very large fields as gravity and solar radiation. His researches convinced him that his L-fields were not just effects of changes that were already taking place in matter (as a feverish rise in body temperature is the result of an illness already present) but in fact could actually cause changes. He concluded that the L-fields both determined and were determined by the biological matter they are associated with. Despite his rigorous adherence to acceptable scientific methods, Burr's work did not receive much attention.

More recently, there are reports that Robert Becker at Upstate Medical Center in Syracuse, N.Y., has demonstrated that mammals may possess the capacity for limb regeneration if an amputated site is stimulated by electricity. He has also uncovered what he believes is evidence of a biological "regulatory field" analogous to fields used to regulate circuits in solid-state electronic mechanisms. These fields, Becker points out, are not within the nervous system. They reside somewhere outside it.

It is unclear whether the fields Rupert Sheldrake proposes are related to Becker's fields, L-fields, or Kirlian fields, but in theory Sheldrake's morphogenetic fields have similar properties—and some surprisingly different ones. It is particularly these different properties which make Sheldrake's a looking-glass theory.

14. The Habits of Matter

Like Bohm, Prigogine, Jantsch, and the Bootstrappers, Sheldrake decided there were definite limits to the mechanistic approach to nature. Beyond these limits, something else is happening. He sensed there must be life fields of some sort which give creatures their form and movement. But the field theories that had been proposed by the vitalists were, he realized, too vague and metaphysical. If, as Bergson said, such life fields were qualitative, then how could their existence be demonstrated by science? Such concrete fields as those discovered by Burr posed other problems. Where did these fields come from? Were they inherited? If so, by what process? More important, what could be the relationship of such life fields to matter?

A NEW FIELD THEORY

In formulating his answer to these questions, Sheldrake took a daring looking-glass step. He proposed that it is not only *life* which is guided in its form by hidden fields, but also the inanimate world of crystals, molecules, and atoms. Like the other looking-glass scientists, Sheldrake found he had to discard the conventional scientific assumption that the world of inanimate matter evolves at some point and becomes animate. He had to see that what we call inanimate is also, in a sense, alive.

With this step, Sheldrake was able to propose the existence of his morphogenetic fields—hidden fields which give regular shape and movement to the universe. From particle to human to galaxy, all growth and form is determined by action of morphogenetic fields on matter. These fields serve as a channel or blueprint. The formation of an atom from a nucleus and electrons is guided by one field, the shape of a molecule by another, the regulation of a cell by yet another. Each field interlocks with others, and the field of a larger form orchestrates the morphogenetic fields of the smaller forms within it. An individual human being—atoms, molecules, tissue, organs, systems—is composed of literally billions of fields, all directed in an ascending interlocking order up to the general field that is the person.

Consider a dandelion springing up unwanted on the front lawn. Its shape and growth have been directed from seed to flower by a morphogenetic field (actually a web of fields) for that species of dandelion. It isn't just that particular dande-

lion which is governed by the field. The field also is guiding all
the other dandelions of that species on the lawn, and indeed
every member of that species of dandelions all over the world.
Not only that, this field has also governed all dandelions since
the beginning of dandelion creation!

Sheldrake's morphogenetic fields are unlike any field
biologists have proposed or physicists have discovered, because they do not obey our current laws of space and time.

At first look, they may seem like Platonic archetypes floating in an eternal, abstract realm. Plato had said that in a
dimension he called "reality" ideal forms existed and that the
forms we see every day are copies of these (imperfect copies,
unfortunately). But Sheldrake's fields are not ethereal
templates which stamp their eternal shapes into matter. They
are looking-glass fields. *They are themselves formed by the very
things they're forming.*

Over time, the morphogenetic field of a dandelion species
doesn't remain exactly the same: The field is continually subtly modified by every dandelion that exists or has ever
existed. The experience of the individual dandelion as it develops in the pressures of its environment is transmitted to
the field. The field, in turn, transmits this total experience to
the form of every dandelion. Adjustments a species has to
make in order to live in one location (such as the adjustment
the peppered moth made to pollution) modify the field and
create a tendency for similar adjustments in an entirely different geographic location if the environment there is similar.
Neo-Darwinians, of course, would claim that this explanation
is superfluous. According to evolution, similar environments
would naturally produce similar adaptations. Yes, Sheldrake
says, but the adaptations in different locations take place
more quickly than could be the case if random mutation and
selection were having to start all over again at a new site. The
more individuals of a species or variety that occur, the more
the field is reinforced. The strengthened field, in turn, makes
it more likely for that species or variety to appear. Some fields
have been around so long and have been reinforced so often
by events that they're effectively changeless. For example, as
more and more energy took the shape of the hydrogen atom,
the hydrogen atom field was reinforced. This field, in turn,
made hydrogen atoms more likely to occur. At this point in
time, the hydrogen atom field is so strong that hydrogen
atoms occur as a law of nature and there is virtually no dif-

ference between one such atom and another. This also applies to hydrogen bonding with oxygen to form the water molecule. The field of this molecule is by now very powerful and causes water molecules to form all the time, throughout the universe.

Sheldrake's fields thus depict a universe in which laws of nature—such as the laws that form atoms and bond atoms into molecules—are built up. Laws are, in effect, habits, reinforced by repetition. Like other looking-glass scientists, though with a new twist, Sheldrake believes the laws of nature aren't eternal, they evolve.

Sheldrake calls the process by which forms in different times and places affect one another through participation in the field "morphic resonance." Suppose that a number of violin strings are stretched on a board and one of them is plucked. If one of the others happens to have exactly the same characteristics of tension, mass, and length, then it too will begin to vibrate without being touched. Strings that are "in tune," or vibrating at the same frequency, will transfer energy to each other through resonance.

Sheldrake believes that similar forms resonate and reinforce each other. When one variety of dandelion succeeds in an environment, it sets up a vibration. Dandelions in a distant place with a similar environmental and genetic tension begin to resonate. The stronger the resonance, the more likely that new generations of dandelions will pick up the vibration and take that form. But the word "resonance" is used only metaphorically. Unlike Burr's L-fields, Sheldrake's morphogenetic fields are not transmitted *energy*. They aren't physical fields at all. They exist in another, nonphysical, dimension. They can't be measured directly by a dial or counter, though, as we'll see, Sheldrake has proposed ways of detecting them that might lead to indirect measurements. Unlike physical fields, they do not diminish with distance. They enter into time and are affected by time, but once a field comes into existence it doesn't die, though the objects or creatures it helps shape may become extinct.

Morphic resonance takes place when a morphogenetic field attaches to a basic unit such as an atom, molecule, or cell. Sheldrake calls this unit a "morphic germ." Once attached to the morphic germ, the field guides other atoms, molecules, or cells into their correct positions, and when this is accomplished, it has created a new, more complex germ. A "higher" field attaches to this germ and comes into play until the overall field that is that organism or form is accomplished. In the case of biological forms, the succession of fields work *in conjunction* with the DNA to guide cell growth and direction. They are like a television signal acting on the circuitry of a TV set—the two together creating an animated form. In the case of inorganic forms, the fields work in conjunction with the forces and properties of matter. Once an entity has its final form, the morphic field remains in place and stabilizes it against fluctuations in the environment.*

So despite the fact that when you meet a friend after six months not one molecule in that friend's face is the same, the face has kept the same basic shape thanks to your friend's morphogenetic field.

The morphic field, therefore, helps replace lost parts (regeneration); it directs the growth of the acorn into the oak. In fact, the existence of a morphic field would answer all the questions of forming Sheldrake posed. In regulation, for example, a sea urchin embryo cell may be taken away but the embryo still develops into a whole, if smaller, sea urchin. This is possible because the morphic field directs the process, though it has less matter to work with (so the urchin is smaller). In reproduction, the sperm and egg create a morphic germ to which a field attaches and begins the process of unfolding an organism.

Morphogenetic fields could also provide an explanation for why transplants work. As Bohm pointed out, the usual view of this has behind it (in subtle forms) the philosophy that the universe is made up of parts, so parts are interchangeable—a body is in principle no different from a car. Sheldrake's hypothesis offers a holistic alternative. An organ from another person or even an artificial organ can be integrated into the fantastic complexity of the body because the morphogenetic field (interlocked fields) acts to close the gaps and smooth the transitions. The field is the instrument of the body's wholeness. Presumably the more transplants are per-

---

* A Cambridge researcher, Michael Bates, has recently claimed discovery of "guidepost" cells which direct growing nerves into their proper places among the labyrinths of other cells. One feature of these guideposts is that each one attracts the growth cone of the developing nerve cell only up to a point, after which another guidepost is needed. Bates' description sounds similar to Sheldrake's idea of a morphic germ attracting growth up to a certain moment at which a new germ takes over.

formed, the more readily bodies will accept these exchanges, because the morphic field will be modified to "expect them."

Sheldrake argues his fields can also be used to explain habitual movement. He calls fields which govern movement "motor fields."

Animal movement in feeding, reproduction, growth, blood circulation, digestion, etc. is controlled by a hierarchy of motor fields, and these fields also affect behavior and instinct. Even complex animal behavior, communication, and social interactions are controlled by motor fields. This is particularly intriguing when it comes to organizations of ants, bees, wasps, and termites in which genetically identical individuals perform quite different roles. If the society is controlled by a motor field, it would explain how small insects can perform quite new tasks in perfect coordination and alter their role within the society when necessary.

In *African Genesis*, writer Robert Ardrey describes a sight which would be very well explained by the existence of motor fields. What he saw was a colony of flattid bugs. Flattid bugs come in different shades; most are coral, at least one in every egg batch is green, and several are in-between shades. The colony arranges itself in the shape of a flower. "I looked closely," Ardrey wrote. "At the tip of the insect flower was a green bud. Behind it were half a dozen partially matured blossoms showing only strains of coral. Behind these on the twig crouched the full strength of flattid bug society; all with wings of purest coral. . . ." His colleague, famed anthropologist Louis Leakey, stirred the colony with a stick. The bugs fluttered around in the air. "Then they returned to their twig. They alighted in no particular order and for an instant, the twig was alive with little creatures climbing over each other's shoulders in what seemed to be random movement. . . . Shortly the twig was still and one beheld again the flower."[1]

If an experimenter pokes a hole in the pot made by a potter's wasp, the insect quickly reseals it. This is a "new" behavior, in effect, because it is an adjustment to unusual circumstances. Just as fields which guide form move the organism or entity to complete its whole shape, so motor fields move the wasp toward the wholeness of completing its goals. Sheldrake says flexibility is part of wholeness. If it takes a new response such as accepting an artificial organ or sealing an unusual damage to the nest, wholeness will find the way.

According to orthodox biological theory, behavior is either innate or learned. Innate behavior is embedded in the genetic code, while learned behavior appears as subtle changes in the nervous system. Learned behavior can never be passed on to offspring except through teaching. Sheldrake overturns this distinction. He sees behavior as coming from three causes: genetic inheritance, morphogenetic fields which control the overall development of the nervous system, and motor fields which shape the behavior patterns of similar animals.

According to Sheldrake, it is perfectly possible for behavior learned by one animal to be passed on to others without any direct contact. For example, an inbred strain of rabbits can be trained to respond to a particular stimulus. This experiment is repeated over and over again until the motor field is reinforced. At this point, geographically remote but genetically similar animals will be found to learn this behavior more easily.

Sheldrake suggests that even human behavior is influenced and aided by motor fields. There are fields associated with cooking, toolmaking, hunting, farming, and weaving—activities that have been performed over and over again for thousands of years. In addition there are historically new tasks such as car driving, piano playing, running the four-minute mile, and flying a jet, which, according to Sheldrake, should become generally easier and easier to learn as time goes on. Sheldrake even suggests that under carefully controlled conditions it should be possible to show experimentally how successive humans learn such tasks more and more quickly. (Of course, as one wag pointed out, despite all the children learning new math, it doesn't seem to have gotten any easier.)

In a similar way, Sheldrake argues that acquired physical characteristics such as a camel's knee calluses can be passed on by generations of kneeling camels, creating and reinforcing the field which covers callus development. Thus Sheldrake gives new meaning to Lamarckian theory. It offers a mechanism for the transmission of the individual's experience with his world to future generations. Lamarck's theory was defeated because it required an acquired characteristic like calluses or a learned ability such as walking a high wire to be genetically transmitted directly to individual offspring. Scientists argued that no direct feedback between such characteristics and genetic germ cells in the animal's reproductive system was possible. Sheldrake's innovation is to suggest that calluses or high-wire walking could be transmitted generally to all—though in the case of high-wire walking it might take a million years of acrobatics for the general population to feel the effect. In Sheldrake's formulation, the individual's influence is not a simple cause-effect link to its offspring but to the field of the entire population via the field, that is *through the whole*—obviously a looking-glass perception.

In answer to molecular biologists and mechanists who claim his theory of hidden fields is totally unnecessary to explain form and behavior, Sheldrake offers an analogy. If someone who knew nothing about electromagnetic waves were to investigate a TV set he might first think it contained little people whose images he saw on the screen. When he looked inside and found transistors and tubes he might adopt hypothesis similar to that of the reductionists—that the images resulted from some interaction of these mechanical parts. This hypothesis would be supported if he found that by taking out some parts he could distort or destroy the picture. If at this juncture someone (like Sheldrake) were to suggest that the picture did not result from these parts but depended on invisible influences entering into them, the investigator would doubtless reject the idea scornfully. He would argue that the set weighed the same when turned off or turned on. He would admit that he couldn't, just now, explain everything from the interactions of the parts of the box but eventually he was sure he would be able to. Compared to the power from the electrical outlet which drives the TV, the power of a TV signal is very weak and subtle. But it's obviously crucial. Sheldrake's analogy recalls Bohm's analogy of the two TV cameras and the fish. In both cases, looking-glass theorists point to processes hidden away in another dimension beyond the apparent correlations of mechanical parts.

PACKING FOR A TRIP TO THE EVIDENCE

The evidence for Sheldrake's hypothesis is at present exceedingly slight—but dramatic. He claims the evidence is thin because scientists haven't thought about looking for it, not because it isn't there.

After a chemist has created a totally new compound, he will attempt to produce pure crystals of his new substance—provided, of course, that it can exist in crystal form. This process of the first crystallization generally proves to be troublesome. However, once a first crystallization has been achieved, scientists find it is curiously much easier to repeat the experiment a second or third time, and it continues to get easier. The conventional explanation is that tiny crystals from the first crystallization have contaminated the laboratory and fall into the second solution, where they act as "seeds" for crystallization. (Dropping a similar crystal in solution always speeds up crystallization.) But this speeded-up effect has also been observed when the first crystallization takes place at considerable distance from the second. Scientists explain that the crystals must have been carried as contaminants on lab coats or in the hair or beards, or microcrystals are carried into the upper atmosphere. Here the theories of conventional science begin to sound even wilder than Sheldrake's.

According to Sheldrake, the first crystallization is so difficult because no morphogenetic field yet exists to guide crystal formation. But with additional crystallizations this field is reinforced and controls the arrangement of atoms and molecules, so the process is speeded up.

In the realm of living entities, Sheldrake says experiments have already been conducted which inadvertently confirm his hypothesis. In one such experiment a particular genetic strain of mice was trained generation after generation to perform a particular task. By the time fifty generations and twenty years had elapsed the new generations of mice did in fact learn their task faster than their distant forebears. At the time, however, the experiment was judged inconclusive because a strange thing had happened—*the control group also had the ability to learn the task faster.*

Conventional thinking could give no reason for accelerated learning in the control group, so, for this reason, the experiment was judged valueless. However, Sheldrake points out that accelerated learning in both the experimental *and* the control group is to be expected on the basis of motor fields. By strengthening the motor field over fifty generations, any genetically similar mouse—control or otherwise—would be affected. Sheldrake suggests that specifically designed experiments should be performed to confirm this effect using groups of mice located hundreds of miles apart.

He cites similar effects in hybrids. If one of the parents of a hybrid comes from an old line and one from a recent mutation, the hybrid offspring show more of the characteristics from the long-standing variety. If both varieties are of approximately equal age, the characteristics of the offspring will be more or less equally shared. These results are not well explained by a strictly genetic theory but are perfectly comprehensible if one assumes that the more individuals of a type have been grown, the stronger its field will be and the more that field will dominate offspring.

Could Sheldrake's fields also explain why particle physicists find it progressively easier to locate a particle in their accelerators once the first one appears? If so, then the theory offers a curious look at the observer-observed question: The physicist, like a plant breeder, is cultivating a variety of particle. The more he observes it by looking for it with his machine, the more it will exist. In a sense he is abstracting it out of the whole, and his observations participate in establishing it more and more firmly in reality (the explicate order). One of the Bootstrap physicists is reported to be devising experiments to test Sheldrake's hypothesis on this point. The point has an interesting echo of Kuhn's observation that once a paradigm shifts the data begin to change, as did the data on atomic weights once Dalton's theory was accepted.

Sheldrake has himself proposed a number of specific experiments to test his hypothesis. These range from crystal growing to measuring learning rates in rats. By offering such experiments Sheldrake has tried to set his theory in a "falsifiable form." Unlike the earlier theories of the vitalists, his can be subjected to experimentation. The field itself can't be measured directly, but its effects can be detected and possibly

indirectly measured by, for example, comparing population size to learning speed and then developing formulae.

Yet the theory's orthodox scientific form didn't keep it from being received by a flurry of scientific outrage. The influential British science journal *Nature* condemned Sheldrake's book as a "candidate for burning." Scientists who would probably have criticized the hypothesis severely if it hadn't been in falsifiable form charged that falsifiability isn't enough, claiming a theory has to explain all of its ramifications to be seriously considered. Nobel Prize–winning physicist Brian Josephson pointed out that if all theories were required to meet such a test, very few would survive—and that among the first victims would be relativity and quantum theory. A June 1981 issue of *New Scientist* told its readers:

To absorb what he [Sheldrake] says involves what Thomas Kuhn termed a "paradigm shift," which means putting aside our assumptions on how the world works. This is an uncomfortable thing to do. However, the notion of modern mechanistic science that we have indeed identified all the major forces and fields at work is astonishingly presumptuous. . . .

Sheldrake observes that most molecular biologists don't react violently to his theory but consider it with a smile. "The mechanists are so sure they're right, they're not worried." All of the looking-glass theories share the problem of evidence, Sheldrake's especially. Kuhn has indicated the paradox: A theory has to be accepted *before* substantial evidence for it comes in. It has to be accepted because it is a novel way of looking and scientists will have to put on the new glasses in order to see anything. Fortunately for Sheldrake, at least some tests will be conducted. A $10,000 prize for experimental tests has been established by an independent organization and *New Scientist* has held a competition for the best test design for Sheldrake's hypothesis.

The winner of the latter contest was an ingenious idea. Nursery rhymes have been learned by hundreds of thousands of young children for generation upon generation. According to the Sheldrake hypothesis these particular word patterns should now be particularly simple to learn. Suppose therefore a Japanese nursery rhyme is presented to an English speaker in the form of phonic symbols, along with a similar nonsense rhyme. The morphic field associated with the long-established rhyme pattern should make it much easier to learn than the control rhyme. Such an experimental test would have to be performed under strictly controlled conditions but would not be particularly expensive to carry out.

If Sheldrake's theory is corroborated by experiment it will inevitably open many questions and lines of conjecture.

MAKING CONNECTIONS

In 1982, Sheldrake met with David Bohm on several occasions and the two scientists began exploring the connections between their theories. At this point they agree that morphogenetic fields can be considered aspects of the implicate order. They have begun the process of elaborating this relationship.

In an obvious way, Sheldrake's theory appears holographic: If each individual dandelion contains atoms and molecules governed by fields that also affect the atoms and molecules in stones, people, and planets, then each form is implied in every other—through the common resonance of fields. The two scientists have, however, considered a much subtler connection.

We'll explain the idea which they've come up with so far by overlapping two images.

For the first image you'll need to recall Bohm's and de Broglie's theory of the guide wave (see page 140). Remember that this is the wave which Bohm thinks governs the whole quantum experiment. It is like a radar beam directing an airplane into position. This guide wave acts nonlocally and is responsible for guiding the particles in the double-slit experiment into their patterns. In a similar (or identical) way the morphogenetic field guides atoms or cells into place to form a structure. Bohm believes the guide wave (and hence the morphogenetic field) is actually a very "subtle" form of energy. It is energy from the implicate order and so exists in the multidimensional reality beyond our three-dimensional space and time. It is subtle but, because it is multidimensional, it is powerful. Remember the fishtank analogy? Compare the two-dimensional television fish to the three-dimensional solid fish in its tank. Implicate energy is the real fish.

The guide wave of implicate energy is a *formative wave*, Bohm says. It is a wave that forms things (like particle patterns). So it is a morphogenetic wave or "cosmic memory." This cosmic memory wave is a relatively autonomous subtotality in the implicate order.

The second image involves time. In Bohm's implicate order the present moment in which you are reading these words can be viewed as an explicate projection from the higher dimension of the whole. This projection then folds back into the whole like the dye drop retreating. Another present moment immediately takes its place, and this, in turn, reenters the whole. As your eyes cross the page, each moment is related to the previous moment through the whole. Thus the present moment is a subtle reflection of all the past. In turn, whatever happens in the present will, as it reenfolds into the implicate, affect the whole and therefore will affect how the future unfolds. This is analogous (or identical) to the picture of morphogenetic fields affecting the dandelion—and the dandelion's experience with its environment, in turn, modifying the morphogenetic field.

Now we'll try to overlap the two images. The guide wave or cosmic memory—that subtle nonlocal energy emanating from the multidimensional implicate—acts on the present moment. Then it acts on the next present moment, and the next. It acts to give a shape to the succession of moments. It guides whatever is unfolding in those moments of space-time. If what's unfolding in space-time is cells, it guides them into place to form a dandelion. Once the dandelion is formed, the guide wave remains in place to give it continued shape as new moments unfold. But as these moments fold again into the whole, they carry an imprint or afterimage of energy back into the implicate, where they affect the guide wave and subtly change the cosmic memory.

All this works, of course, because implicate and explicate, morphogenetic form and the entity it shapes, aren't really separate—they're different dimensions of the same thing.

The two scientists have also discussed a primary question which Sheldrake doesn't attempt to answer with his theory: How does a new form *first* come into being, before any morphogenetic field exists?

The neo-Darwinian answer is that the first appearance of a new variety or species is the result of random genetic mutations. Sheldrake says that's only one possible explanation—there are others. The appearance of new forms could also be the result of some kind of creative principle as yet unknown but inherent in life or in the universe. Or there could be a conscious agent or transcendent reality overseeing the whole universe which (or Who) creates new forms. Sheldrake has stipulated he isn't prepared to say which of these or any other ideas explains the first appearance of a new form or totally new behavior. He believes a line should be drawn between what science can know by observation and experiment and what it can only speculate about. Neo-Darwinian randomism is, he says, as speculative an answer as God.

From his perspective, Bohm is prepared, however, to combine morphogenetic fields and the implicate order and answer the question of creativity. He compares the appearance of a new biological form to a flash of insight by the universe. In an insight "you realize the whole in the mind." Similarly, it is the movement of the whole universe which creates a new form. The form then becomes established and strengthened by the processes Sheldrake describes. Because both forms and the formative field are not fixed things but must be constantly re-created, it is almost inevitable for creative jumps to occur. In a sense, recurring and habitual forms make creativity possible. Of course, where all this wholeness comes from is another question. Or maybe it isn't a question, for in a universe in which everything affects everything else, what would such a question mean?

Both the Bohm and Sheldrake hypotheses agree on the unity of human consciousness. Consciousness as a whole is a morphogenetic field giving a general shape to each individual's consciousness. Each individual consciousness also forms its own field, including its experiences and memories. This individual field resonates and modifies the field of human consciousness as a whole, affecting the future. The ethical-moral-psychological implications of such an idea are obviously enormous. It makes the individual responsible for the whole. Among many other possibilities, it could lead to investigations of particular cultural traits and individual traits as the building up of habit fields. The "laws" of how such fields evolve and interact might be defined—what Bohm calls the "laws of relative autonomies."

Like Bohm's theory, Sheldrake's offers a way to explain and explore such disputed phenomena as archetypes of collective consciousness and psychic transmission. There is also some apparent connection of Sheldrake's hypothesis to the Prigogine paradigm. This can perhaps best be illustrated with an anecdote told by Lyall Watson in *Lifetide: The Biology of Consciousness.*

When a monkey tribe on an island near Japan was introduced to freshly dug sweet potatoes, the monkeys, whose other foods required no preparation, were reluctant to eat the dirty potatoes. One female discovered they could be washed. This discovery was monumental, like discovering the wheel. The female taught this behavior to a few of the troop, who in turn taught it to some others. But most of the monkeys did not learn it. Then, quite suddenly, the behavior seemed to become universal. Watson writes:

Let us say, for argument's sake, that the number [of potato washers] was 99 and that at 11 o'clock on a Tuesday morning, one further convert was added to the fold in the usual way. But the addition of the hundredth monkey apparently carried the number across some sort of threshold, pushing it through a kind of *critical mass*, because by that evening almost everyone in the colony was doing it. Not only that, but the habit seems to have jumped natural barriers and to have appeared spontaneously, like glycerine crystals in sealed laboratory jars, in colonies on other islands and on the mainland in a troop in Takasakiyama. [Emphasis added.][46*]

The anecdote tantalizingly suggests that the appearance of spontaneous order described by Prigogine's theory might have something to do with nonlocal fields described by Sheldrake. Is there a critical mass necessary before a morphic field can form? Could morphic fields explain how dissipative/autopoietic structures remain stable and proliferate over time? The nature and extent of this connection remains to be explored. Certainly it would be interesting to know if there is a relationship among the fields of different entities and whether these are governed by the principles of co-evolution. How are the fields governing the bacteria and the fields governing the Gaia system related? Are they related like the fields of a cell and a body? Do they reinforce each other? Arise together?

It seems evident that Sheldrake's hypothesis will, if it stands its tests, need expansion by some larger theory such as the implicate order or co-evolution. A larger dimension would

---

* It should be noted that the paper which Watson cites as documentation for this phenomenon does not show a sudden jump in the monkey learning rate. Watson says the real evidence for the phenomena exists in anecdotal accounts by scientists who are reluctant to face the implications of their discovery and have (evidently) unconsciously structured the data or the experiment to fit the currently dominant paradigm.

be required to answer, for example, how it is that dinosaurs could be extinct. Dinosaurs prowled the earth for millions of years. Presumably their morphic fields were strong and well reinforced. Nevertheless, they suddenly vanished, possibly the result of a catastrophe like the impact of an asteroid. Sheldrake would say their fields still exist, but the dinosaurs do not. In the context of the implicate order this disappearance is comprehensible. It is also comprehensible in the co-evolution paradigm. In both cases, the dinosaur fields would be seen as part of a larger whole unfolding movement of many interpenetrating fields—fields including the ecosystem and patterns of climate. Dinosaurs didn't vanish, they became part of the reordering into a higher complexity, or their ensemble was dispersed again and wound into different forms in the implicate whole. Speculation can easily run wild. Such openness may be the mark of a fertile paradigm or of an illusion.

Sheldrake's theory is young, untested, fragile, and provocative. Whatever its fate, we have seen it springs from a looking-glass perception and contains most, if not all, of our looking-glass ingredients.

Rupert Sheldrake

5

Karl Pribram and the Looking-Glass Mind

"_Now_ cut it up," said the Lion.

15. Search for the Engram

At one end of the Schrödinger's cat experiment, a radioactive atom decays and its particle is emitted according to the statistical laws of the quantum. As he tried to explain this end, David Bohm wandered into the mysteries of the implicate order. In the middle of the experiment, the cat stretches and yawns, unaware of its fate. Lying there sleepily, a complex of chemical reactions, this constantly changing yet miraculously stable form defies the march of entropy. In an effort to understand this fur-coated violation of equilibrium, Ilya Prigogine and Erich Jantsch wound themselves into the labyrinths of order through fluctuation and co-evolution, and Rupert Sheldrake entered a realm beyond time and space.

At the near end of the experiment we now see another figure, curiously opening the box lid to peer inside. In conventional theories he has been somewhat separated from the first two elements of the experiment, isolated by chance and by the belief that he is an independent part of the whole. In our looking-glass theories, however, he has been implied all along. His expression tells us that he is thinking, remembering, correlating as he looks inside. We can conclude by these movements that he is "the observer."

Our fourth looking-glass expedition will set out to see what is in this observer's mind—how his observing is possible and what observing is. In the process we will find that in the looking glass universe this end of the experiment indeed reflects the other end, and everything in between. There are not three elements in the experiment after all—just one.

MODELS

In 1921 the world's first automatic telephone exchange opened in Omaha, Nebraska. At last subscribers could reach a number anywhere in the city simply by lifting the telephone and dialing. Electromagnetic relays inside the telephone office were triggered by the dial signal and connected the incoming call to one of the thousands of lines that made the city a living organism. Omaha had become a complex web of wire nerves that reached into every city block.

When it came to brain science the metaphor could be reversed. In early models the brain was viewed as a telephone exchange with nerves like telephone lines. If a person accidentally touched a hot stove messages from the hand were pictured as being sent along a nerve pathway into the skull. Within the brain a form of switching took place to direct a new signal along a second pathway to the muscles of the arm. The muscles contracted and the hand pulled away from the stove.

As far as it went, the "brain as telephone exchange" was not a bad idea. It stimulated the imagination and enabled mathematical techniques developed in communications theory to be applied to the nervous system.

Eventually, however, it became clear that the brain is *not* a telephone exchange and nerves are *not* telephone lines. For one thing, the way in which signals are carried by nerve fibers is very different from electrical signals in a telephone or telegraph wire. More important, the brain and nervous system are far richer and more complex and employ far more processes than any telephone exhange could even be conceived of doing.

The practice of representing the brain by some artificial model is nothing new. In the eighteenth century the leisured class was given to the fashion for automata—fountains that played in complex patterns determined by hydraulic links, or life-size manikins that could write or play a musical instrument. What could be more natural than to model the activity of the brain on such things?

As we have seen from the beginning of this book, while metaphors may offer a temporary insight into a complex system, they are necessarily limited, sometimes gravely so. A model is never the thing it stands for. Nevertheless, models and metaphors have a tendency to assume a life of their own.

In our own century the computer has replaced the telephone exchange as a model for the brain. Electromagnetic relays have been replaced by data processing, programs, and memory storage as elements of the metaphor. Human behavior is treated in terms of heuristic learning systems, vision in terms of pattern recognition and image compression.

In the mid-1960s a new metaphor emerged to stand beside that of the computer brain. Thanks to the energies of neuroscientist Karl Pribram, the "holographic brain" provided a new set of insights into processes which take place inside the brains of animals and humans. The value of this new metaphor is that it requires a new way of thinking, a new approach to the problem. Yet in the end it will doubtless prove as limited as any other model.

Attempting to understand the brain is one of the most exciting and difficult problems that face science at the end of this century; it is, in the view of some, "the final frontier." The appearance of the brain in evolution gave to the universe the possibility of self-realization. The brain is the organ with which the universe knows and perceives itself. But this loop of self-reflection threatens to engulf the observing brain in a paradox. What is the relationship between the physical brain and what appears to be quite nonphysical—awareness and the human mind?

Within the human mind is a range of experience and feeling—emotions, both violent and refined, dreams and desires, invention and intuition, creativity and destruction, the ability to manipulate such abstractions as music and mathematics and the words on this page, and the capacity to be carried away by ideas. Mind also includes the idea of "self," a feeling of identity that persists over time. Even more mysterious is the dissolution of self that occurs during creative work or transcendental experiences. Above all, the alert, inquiring mind can direct attention to its own nature and to that organ within the skull which medical science has concluded is the seat of awareness—the brain. It is clear that we can't scientifically answer the question of the relationship of mind and brain until we understand this organ.

In the case of animals, the neuroscientist feels free to probe the inner structure of brains, monitor electrical signals, and explore the effects of selective damage. With humans this isn't possible, because we have decided that human brains are more sacrosanct. Until the recent invention of CAT, PET, NMR, and ultrasound scans, examination of the human brain was therefore restricted to autopsy and cases of surgery or accident. Evidence about the correlations between behavior or awareness and the human brain came through diagnosis.

Diagnosis involves such elements as general appearance, history, the nature of the patient's pain and dysfunction, pulse, eyes, tongue, the feel of the skin, the sound of the tapped body, and the external feel of internal organs. Just as someone recognizes the face of a friend, so a doctor recognizes the face of a disease or injury. It is the success of this technique in the case of brain injury, disease, or tumor and, more recently, drug impairment that built up a picture of the relationship between the brain and behavior. Ironically, human destruction caused by the First World War produced rapid advances in brain science. High-speed rifle bullets caused localized damage to the brain and made it possible to map brain function. Accumulating diagnostic data enabled doctors to locate within the brain the sites of hearing, vision, control over particular limbs, and even more subtle correlates such as drives and emotions.

Gradually a picture was built up in which the brain was seen as something like a factory with each machine or process housed in a given location. If the factory was damaged, certain processes would stop while others carried on unaffected. This correlation between brain and behavior received a considerable boost in the early decades of this century through the work of Wilder Penfield at the Montreal Neurological Institute. Before conducting brain surgery, Penfield would obtain his patient's consent to perform harmless experiments on the exposed brain tissue. Using a small electrical probe, the neurosurgeon selectively stimulated tiny regions of the brain and observed the results. In some regions the patient would twitch an eye or raise a finger. In others he would experience a strong smell or recall a sudden memory with an intensity that approached that of a hallucination.

Penfield's research suggested that memory, like other functions, had a specific location within the brain. Memory was like the factory's card index file, with each item written on a particular card. Recollection was simply a matter of going to the file and extracting a particular card. Since associations can stimulate memory, the index itself would be cross-referenced in quite a complex way. In certain of Penfield's experiments, a cell or group of cells was stimulated and the memory cards they contained were projected into awareness.

THOSE ELUSIVE LOCATIONS

This faith in the localization of memory was badly shaken by the work of Karl Spenser Lashley. Lashley performed delicate

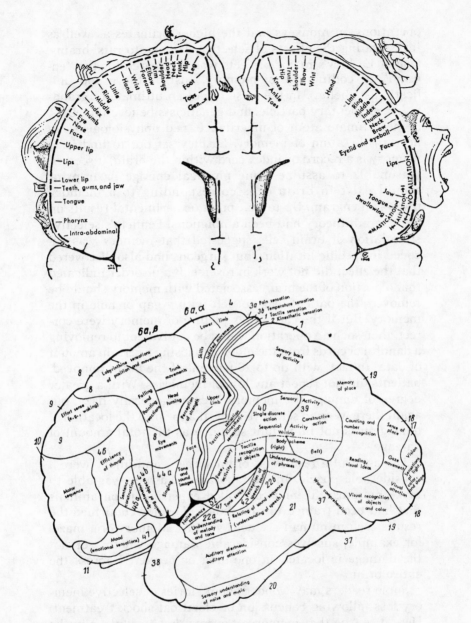

The upper diagram is a cross section through the center of the brain. Regions corresponding to different parts of the body and its functions were very precisely located by surgeons. The lower diagram shows an attempt to localize subtler brain functions in the belief that everything in mental life must have its particular site in the brain. Such maps were invalidated by discoveries that many brain functions, including memory, must be somehow spread out or "delocalized" across the brain.

operations on monkeys and the higher primates as well as investigating the effects of selective damage to rats' brains. One of Lashley's major projects was the search for the "engram" or coding site of memory. Just as the grand unificationists believe in the existence of an ultimate or grandfather elementary particle, and geneticists believe that DNA is the ultimate atom of inheritance, neurophysiologists believe in the atom of memory. Lashley set out to find it. If memory is a record or index card within the brain, it seemed reasonable to assume some physical change or imprint should exist in brain cells corresponding to a memory trace—an engram. Up to that point, experimental results on the engram theory had been ambiguous. Penfield's selective stimulation of brain cells suggested that memory was encoded in specific locations, but surgeons had also discovered that the effect did not work in reverse. If a disorder indicated that a portion of the brain associated with memory should be removed, the patient was not left with a gap or hole in the memory. But if the card-index picture of memory were correct, then such an operation would be equivalent to removing a handful of cards from the index and losing a certain amount of data. In fact, with up to 20 percent of the brain removed, patients did not report any loss of memories. With a greater degree of damage the memory became generally hazy, but again there were no selective gaps. It appeared the location of damage was not important so much as the total amount of damage.

The anecdotal results on brain surgery and injury were a considerable puzzle. In his laboratory Lashley was able to make more careful and controlled experiments on animals that had been trained to run a maze. Lashley confirmed the results seen in human accident victims. Memory of a maze, for example, survives considerable damage whether or not the damage is located in one area or peppered across the entire brain.

Some readers may wonder about stories of selective memory loss following concussion or electrical shock treatment. Here it is true that memory of events that took place in the several hours or minutes before an accident may be lost. The reason is not fully known but probably involves two different forms of memory storage—long-term and short-term. Long-term memories are the type studied by Penfield and Lashley. They stretch back into childhood and we may not be directly aware of them. By contrast, short-term or "active memory" operates for such things as a telephone number recently read out of a book. They are transitory memories of which we are directly aware.

There appears to be a two-way traffic between short-term and long-term memories. Some short-term memories are presumably sent to permanent storage in long-term memory. When we later recall something we remove it from long-term storage and transfer it to active (short-term) memory. At this point we become aware of the memory. After a shock or during temporary amnesia, some disturbance in the transfers between active and long-term memory takes place. Active memories may be wiped out, or there may be a disruption in traffic between active and long-term memory so that immediate events are not stored and recent events are not recalled. According to this theory, which owes a great deal to the computer model of the brain, temporary memory loss would have nothing to do with the permanent loss of memory that Lashley was searching for.

Lashley also directed his attention to the visual system. He discovered that with 80 percent of its visual cortex removed a rat could still respond to visual clues. In cats, up to 98 percent of the optic nerve could be severed and vision retained. When Lashley's work was combined with that of other laboratories, it appeared that a theory of memory encoded in specific locations just did not work. Somehow memory must be "delocalized" in the brain. Even the maps of brain function versus behavior were shown to be only approximations. While it is generally true that motor ability, say, is governed by a specific group of cells in the brain, in certain clinical cases the situation becomes more flexible. Children who have received severe brain damage as a result of accidents sometimes recover with only minimal disabilities. Even when the speech centers are badly damaged, children may leave the hospital speaking normally. This is reminiscent of the process of "regulation" studied by Sheldrake whereby part of an early form, such as a sea urchin embryo, could be damaged and still develop into a complete entity.

The ability to transfer learning from one limb to the other also appeared to defy a hypothesis of localized brain function. A right-handed person can wrap the toes of his left foot around a crayon and write, albeit with some difficulty. If a group of cells on the left hemisphere of the brain have carefully learned the art of writing, how on earth can this ability be passed to cells in the opposite hemisphere? Can "learning" pass from one side of the brain to the other?

Some scientists hoped to save the localization theory by suggesting that the brain evolved to protect itself from the results of injury by duplicating every function over and over again so that when one area breaks down another can take over. This is as if each memory were entered onto several sets of cards and each card were filed in a different location. This was one possible explanation but an increasing weight of experimental and medical evidence pointed to some sort of delocalizing processes across large areas of the brain and not separate storage sites for discrete memories. In 1929, Lashley published a major book in which he proposed the "principle of mass action," the idea that certain types of learning involve the whole cerebral cortex. In the same book he also stated the "principle of equipotentiality," a suggestion that when one part of the sensory system is damaged its function can be taken over by any other part. Lashley's ideas, however, were not actively pursued by most researchers, who continued to hold to some modified version of the localization theory.

A member of Lashley's team and his eventual successor as director of the Yerkes laboratory was Karl Pribram. The young Pribram had joined Lashley's group as a neurosurgeon. Following the older man's retirement in 1948, Pribram briefly headed the research team before moving to Yale, where he performed pioneering work on the brain's limbic system, a brain area active in emotional responses. A decade later, Pribram challenged the behaviorist assumption that nerve impulses take the form of a "reflex arc" and showed, instead, that they act like feedback systems, whirlpools in which stimulus and response constantly modify each other.

16. Brain Storms

By the mid-1960s, Karl Pribram was directing a number of important researches. He also pondered his mentor's failure to discover the engram. One day Pribram happened to read an issue of *Scientific American* which contained an account of recent successes in holography. This article triggered Pribram's insight into the nature of memory.

We have met the hologram already. It involves enfolding an optical image over a photographic plate. In conventional photography each part of the plate contains a part of the image. In holography, by contrast, a part broken from the plate contains a record of the whole image. If one damages part of the holograph, one does not lose a part of the image it contains.

The analogy with a distributed memory and its ability to withstand brain damage appealed to Pribram. His next step was to discuss holography with his son, a physicist. The more Pribram looked into the basic theory the richer the analogy with the brain seemed. The hologram could provide a model not only for memory but for the brain's visual system as well. To understand the reasons for Pribram's enthusiasm we need to return to holography and examine it in greater detail.

GOING THROUGH PHASES

Thomas Young's experiment on light interference—which was one of the early starting points for the voyage into the quantum—assumes a special significance in holography. Interference takes place whenever waves encounter each other. The process of interference applies not only to light waves but to sound in a concert hall, to ripples on a lake, and, as we shall see, to electrical signals in the brain.

Suppose that slow, uniform ripples move in from a large lake toward the shore, passing across where a boat is drifting. The ripples pass the boat leaving a patch of undisturbed water on the opposite side—a ripple-shadow, as it were. Since waves have a tendency to spread out, this area of calm water does not extend too far and some distance away from the boat the waves meet again. It is in this region that interference takes place.

Let us follow two waves, one from the stem and the other from the stern of the boat as they travel toward their meeting point. If both waves travel exactly the same distance, then when they meet they will match crest for crest. Such waves are said to be in phase. But if one wave has a slightly longer

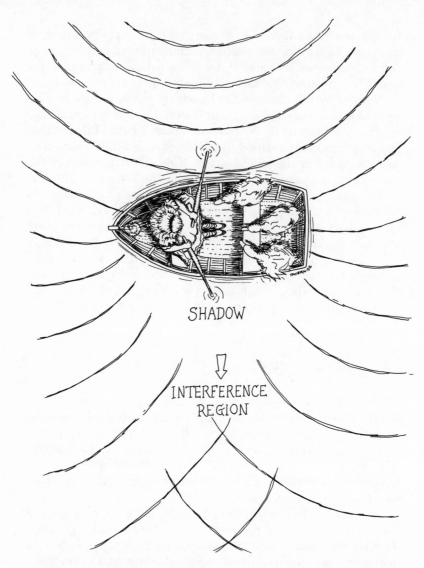

SHADOW

INTERFERENCE
REGION

distance to travel, its neighbor will arrive at the meeting
point first. By the time the crest of the second wave arrives,
the first will have moved on a little, leaving only its trough or
shoulder. Such waves are said to be out of phase. For example, when trough and crest occur together, the waves are 180
degrees out of phase. If two crests meet, the waves are in
phase, or put another way, 0 degrees or 360 degrees out of
phase. As waves meet behind the boat, their individual phases
will be considerably mixed up.

When two crests meet in phase the net effect will be to raise the water. At another point two troughs may meet in phase and cause a deep depression. In some other region the waves may be 180 degrees out of phase so that a crest and trough meet and cancel exactly. Other waves meet in different degrees out of phase. The result is to produce a rich interference pattern of disturbances. This pattern is produced by waves which have traveled from both ends the boat. Mathematically this pattern can be expressed in terms of the waves' phase differences.

The top diagram shows the effect when two waves meet in phase. The second diagram shows waves meeting exactly 180 degrees out of phase. The bottom shows waves meeting slightly out of phase.

Phase and interference also appear in a concert hall and a stereo set, in which sound waves replace ripples on the surface of the lake. In a large concert hall, sound reaches the listener from many different points, directly from the instruments dotted about the stage and by reflection from the walls and ceiling. Since our ears are separated in space, sounds will reach one ear a fraction of a second later or earlier than the other. The sound of each instrument therefore reaches the brain as a signal from both ears with a slight phase difference between the right and left ear. These phases are analyzed by the brain and used to place the sound in space. A sound with no phase difference will be placed centrally. One with a large phase difference will be placed on the left, say, while a sound with an opposite phase difference will be placed on the right.

If these sounds are recorded by a single microphone and relayed through a speaker, only the pitch and loudness of the music will be produced. Phase information is lost. But if the sounds are recorded by a pair of microphones separated in space and then played through stereo speakers, the illusion of sound distributed in space is produced. A sound fed to one speaker will have a fraction-of-a-second difference from that fed to the other and the result produces waves that are out of phase and will interfere in the listening room. The net effect is a simulation of sound spread out in space.

It is possible to play an interesting trick with a stereo system by reversing the wires on one of the speakers. This switches all the phase relationships 180 degrees so that central sounds are exactly out of phase. Two waves from an instrument central on the stage should in fact meet in phase in the room and add constructively. After wires are switched, however, the waves will subtract from each other—crest being canceled by trough. There now seems to be a "hole" in the center of the listening field. If the wires on the second speaker are also switched and the phase is rotated by another 180 degrees, we will have made a full circle and arrived back at the starting point. With both wires reversed the effect is identical to leaving them both alone.

All of this points up the fact that the phase relationships between waves and their interference carry valuable information about the whole or distributed state of things such as the distributed sounds in a concert hall. During optical interference, waves from all over an object meet. In the optical hologram information related to the phases is recorded in the silver halide on the photographic plate as an interference pattern. This pattern encodes information about the whole object and can be used to reconstruct an accurate three-dimensional image of it.

We can see how this works by returning in more detail to the comparison mentioned by Bohm. In normal photography, a lens is used to form an image on a photographic plate. Each point on the object corresponds to a single point on the image and vice versa. In mathematical terms, a transformation has taken place from object to image. With a flat object and flat image this is a particularly simple transformation known as a 1:1 (one-to-one) transformation. In normal photography the only information available is the color of the object and the intensity of light at each point; no phase information is recorded. When a portion of the photographic plate is destroyed, a corresponding portion of the image is lost, with the remainder being preserved intact.

In holography, something quite different occurs. Here light reflected from all over the object is used to produce an interference pattern. The pattern is recorded on a photographic or "holographic" plate. The pattern is not in 1:1 correspondence with the object because the phase information from each region of the object is recorded throughout the holographic plate. Thus, if a portion of the plate is lost, the total image is retained.

There is no single form of holography. Depending on the results needed, a variety of procedures can be adopted. Normally, laser light is used as a source of illumination. Ordinary

Every point of the rabbit doing his balancing act, left, corresponds to a point on the image projected by the lens. It is a 1:1 correspondence.

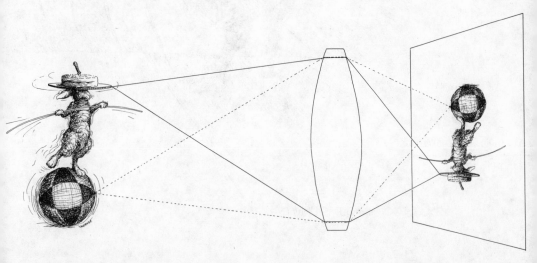

light, from a flame, an incandescent lamp, the sun, or a flashbulb contains waves which are already in a jumble of phases. Such light is not a good form of illumination to use when the recording of retrievable phases is required. In a laser, however all the atoms are triggered to fire at once so the waves of the resulting beam are in phase.

Laser light is split into two beams, one to illuminate the

object and the other to act as a "reference beam." When the two beams meet again, interference occurs. As a result of being scattered by the object, one of these beams contains a mixture of phases, just like the sound from the stereo set. The reference beam contains light which is all in phase. The resulting interference pattern consists of a very fine pattern of light and dark patches, a kind of code.

With the pattern recorded on a photographic plate it is now possible to reproduce the image. This is done by shining the reference beam onto the plate and viewing it from the other side.

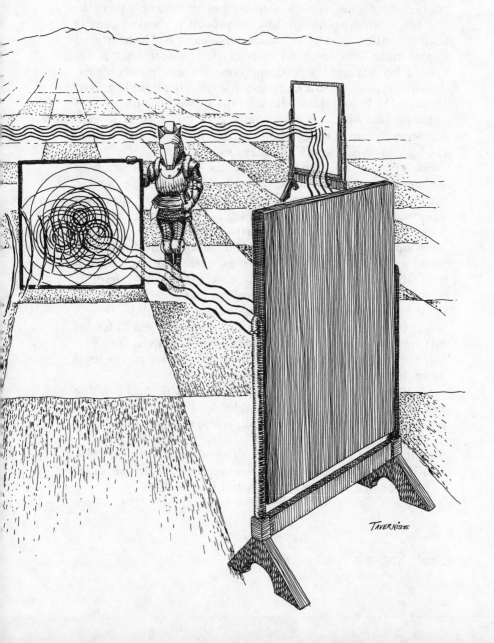

The image appears in three dimensions, and if the viewer moves his head he seems to "see around" the object. The plate itself could be said to contain a record of the reference and scattered beams. When the reference beam is shone on the plate it excites the record and the scattered beam can be seen. It was this very fact that suggested to Pribram a strong analogy between the holograph and the brain's memory storage and retrieval capacity.

A variation of the above holographic method suggested a further analogy to the phenomenon of "association" in memory, where, for example, a memory of a person last seen at a party suddenly calls up the memories of other people there. In the corresponding holographic approach a single beam is used to scatter light from two objects, A and B. The scattered beams meet and form an interference pattern that is recorded. To activate the holograph one of the original objects, A, is illuminated with a laser and this reflected light is passed through the holographic plate. To the viewer, the image of B appears. Likewise if B is illuminated in this way, A is seen. Holography can therefore be used in a form where the presence if one object *recalls* the features of the other.

For Pribram, analogies between the optical hologram in which information about an object is distributed over a wide area and Lashley's idea of the distributed memory were persuasive. When it came to the brain's visual system, another form of holography, "recognition holography," offered additional suggestions. In recognition holography a reference beam and a scattered beam from an object are used to produce the holograph, but in this case a focusing mirror is added. If light scattered from an identical object is then reflected from the mirror onto the holographic plate, then a single bright spot is seen. If the illuminating object differs at all from the original, then no spot is formed. When a photocell is placed behind the holograph, the whole device can be used as an automatic recognition system.

Such a system works by recognizing an object as a whole and not simply by matching individual features piece by piece. Could this be how the brain can spot a face in a crowd or recognize a photograph of a familiar scene? We certainly do not see our friends in terms of eyes, nose, chin but as a totality. Possibly recognition holography could provide clues to how the brain processes its visual information.

SOME "DISTRIBUTED" RESEARCH

The holographic process is certainly suggestive, but to take it any further Pribram had to determine if anything analogous to a holographic transformation actually takes place inside the skull. Particularly, he had to find out if there were mechanisms in the brain which corresponded to the reference and scattered beams and the plate. (More about his successes on that venture in the next chapter.) Pribram's own laboratory work had not been directed toward this holographic theory, however. His contribution to it involved collating and reviewing the experiments of other groups and seeing to what extent they provided support for a holographic model of brain function.

Throughout the late 1960s and into the '70s, Pribram published his arguments in scientific journals. He was not the only scientist to do so, nor could the holographic theory of the brain be said to originate with any one thinker. However, Pribram's support for the theory in scientific publications, lectures, conferences, and popular accounts has gained the theory considerable attention.

Not everyone welcomed it. Some believed that a delocalized memory was produced not by a holographic effect but by electrical fields operating across the brain. Others suggested that the flow of chemical messengers, neurotransmitters, could be involved in the storage of memories and learning. There is evidence to support all these theories. Some of it is persuasive but none of it is clear-cut. Unlike experiments in the physical sciences, experiments involving laboratory animals, perception, and behavior may be capable of a number of different interpretations. Subtle effects can be introduced by, for example, the way a researcher or his assistants handle different groups of animals. When it comes to theories of the brain, it is difficult to apply Karl Popper's requirement that a theory must be cast in clearly falsifiable form. In other words, it is sometimes difficult to *disprove* some biological theories, though they may, in fact, be wrong.

One scientist who first rejected the holographic theory and attempted to disprove it was Paul Pietsch. Pietsch didn't have experience in the field of brain research, but his work on tissue rejection in salamanders had previously convinced him that the body functions by using independent groups of cells.

He was sure that the brain would be no different. In his book *Shufflebrain*, Pietsch describes a series of experiments on the brains of salamanders. This is a curious creature capable of regenerating its limbs following amputation. Pietsch found that the optic nerve of the animals would rejoin after severing and brain transplants would integrate into the brain. Pietsch shuffled or mixed the brains of individual salamanders. Portions of the brain were rotated, hemispheres interchanged. He naturally expected such drastic rearrangements to show up in bizarre behavior, but to his surprise most of the animals continued to behave normally.

Unfortunately for the interpretation of these experiments, salamander behavior mostly consists of eating tubifex worms. Pietsch's experimental animals couldn't be trained to push buttons or run a maze, so it becomes difficult to assess the precise results of changes in this creature's brain. However, the experiments showed that the salamander's feeding mechanisms were able to survive violent disruption of the brain's structure. Pietsch concluded that in some way they are delocalized across the brain.

As a result of his researches, Pietsch was transformed from a skeptic to a convert of holographic brain theory, and he went on to offer some additional speculations about it. In one speculation he suggests that a "distributed brain" is universal—occurring in even the most primitive brains.

Bacteria found in the human gut can respond to very small gradients of chemicals. For example, the minute bacterium *E. coli* will swim toward meat extract and away from alcohol. To do this the bacterium must determine in which direction the concentration of chemical is largest, and this implies that it is able to make an analysis of very tiny differences in concentration across its length. Since the organism is so incredibly tiny, these differences are almost vanishingly small, and some researchers feel that it is impossible for a bacterium to make such an accurate analysis. The alternative hypothesis would be for the bacterium to "remember" the average concentration of a chemical from region to region as it swims along. This idea is even less tolerable to the orthodox, since it implies that a single-celled bacterium can not only store memories but order them as well. How could memory be stored and processed in a bacterium, which has no nerves, let alone a brain? Pietsch believes the hypothesis is reasonable in light of the holographic model. He conjectures that the bacterium's "memory" is in fact distributed holographically all over its surface. By expressing information on chemical concentrations in terms of the geometrical arrangement of certain protein molecules, the outer membrane of *E. coli* becomes an extended "brain." In some transfer process, information stored in the arrangement of the surface proteins is used to activate contractile proteins that cause the flagella (tiny hairs) to move and propel the organism. Pietsch reasons that if the very first primitive memory systems began as a distribution of information across the surface of an animal, then higher stages of evolution could have retained this method of distributed coding.

Henri Bergson, whom Sheldrake studied, speculated that consciousness is not in the brain but is distributed across the body. The controversial psychologist Wilhelm Reich based his therapy on retrieving "memories" which are distributed in the body's muscle tensions.

According to a recent report in *Brain/Mind Bulletin*, a new method of photography has disclosed a bizarre holographic phenomenon in plants. When a hole or square is cut out of a leaf, inside the hole a ghostly miniature image of the entire leaf appears. This effect also seems to echo the ghosts of Sheldrake's morphogenetic fields.

This and other tantalizing bits of research suggest that holographic distribution is basic to many different kinds of order, from photons to "plant consciousness" to human consciousness.

The great excitement of the 1970s in brain research was not, however, holographic theory but the work on the so-called split brain. This work, for a time at least, gave new life to the idea that brain states could be localized. The left hemisphere of the brain in popular and scientific literature became associated with the rational processes of language and logic, the right half with more artistic, intuitive states. Some writers even claimed that the right half of the brain was the seat of creativity. Later research has shown the situation to be much more subtle. As one example, measurements of electrical activity in the brain indicated that the right brain was more active when experimental subjects listened to music, thus supporting localization of nonverbal, intuitive features of consciousness in this side of the brain. However, when musicians listened to music, the *left* brain was more active. It might be argued that this just means that creative artists are more analytical when attending to a work of art, but then we will have the problem of defining creativity. Recent research has also shown the right hemisphere to have analytic capacity.

It now appears that after early childhood each side of the brain begins to specialize in certain activities. But consciousness itself is not composed of parts of activity here and there. It is distributed across the whole brain. Sleepwalkers or people who suffer from "multiple personalities" do not remember things they have said or done when they were in "another state of mind." However, the term "split personality" is a misnomer. The different personalities are not integrated with each other, but each one is in itself a whole, integrated set of gestures, movements, memories. It's something like the hologram. The whole of a person is there but each part (or "personality") displays a different perspective. Shifting the current scientific gestalt away from parts may, in the future, allow a new understanding of wholeness to appear in many areas of research into brain and mind.

17.
Holographic Observer Math

Holograms in the holes of damaged leaves, Pietsch's idea about memory distribution in bacteria, and the notion of holographic structure to the split brain will require much more scientific work before they can be properly assessed. To explore more detailed evidence for the holographic brain, we must return to Pribram.

PATTERNS IN THE BRAIN

The essence of holography lies in the interference patterns on the holographic plate. But how could interference occur in the brain? To answer this question, Pribram turned to a particularly well-studied brain area—the visual system. Anatomists consider the whole system—retina, optic nerve, and visual cortex—as part of the brain. According to this view, in effect, the eye apparatus is a region of the brain projected forward to sit in the orbits of the skull.

When light falls on the back of the eye it stimulates the sensitive rods and cones on the retina and causes them to fire electrical signals. These cells are connected to the visual cortex by means of a complex pathway called the optic nerve. At one time, it was believed that each rod or cone directly stimulated a cell or group of cells in the brain. A square image falling on the retina therefore would stimulate a square set of cells in the brain. This model of vision was similar to an ordinary photograph and involved a 1:1 transformation. It would be difficult for such a model to explain how vision survives damage to the cortex or optic nerve.

More careful work has replaced this early theory and shows that something closer to an expanding ripple of electrical activity reaches the visual cortex. The retina itself is a complex processing unit. Signals from rods and cones are compared to each other and modified before they even reach the optic nerve. The optic nerve is not a single fiber but a multiple system of nerves which stretches from the back of the eye into the brain. All along its length it is traversed by thousands of "local circuit neurons" which interconnect individual nerve fibers. The result is a complex spiderweb of connections which cross the optic nerve again and again. These local circuit neurons do not originate signals of their own but act to enhance or inhibit signals in the optic nerve. At each interconnection, the local circuit neurons compare signals which pass down neighboring fibers of the optic nerve and modify them. A very strong signal will be surrounded by fibers carrying much weaker signals, but farther out, will be surrounded by stronger signals again like ripples in a pond of water. By the time a particular visual signal reaches the visual cortex it will have been modified by tens of thousands of interconnections. No longer can we conceive of a single rod or cone giving rise to a stimulus in a single brain cell. Instead, a ripple of activity spreads out across the cortex. As the signal passes up the optic nerve it becomes modified into concentric rings of activity or spreading ripples.

If several different parts of the retina are stimulated (as they almost always are), then many series of spreading concentric waves travel up the optic nerve into the visual cortex. The final result is that when a complex shape is presented to the eye these spreading waves interfere with each other to produce something suggestive of a holographic pattern. Visual information about an object is distributed by this pattern across the whole cortex.

Evidence for these mechanisms came from careful laboratory experiments on cats, monkeys, rats, and other animals. At the time, most of this work was not done directly as a test of holographic theory, but it has since been shown to be consistent with the idea of holographic interference patterns in the brain. In typical experiments, selected images are flashed onto the eye of an experimental animal and corresponding electrical activity is monitored by probes placed in the brain. Such experiments indicate that the visual holographs are not in fact spread evenly across the entire visual cortex but occur in patches. For every 5 degrees of the visual field a holograph appears to be formed and projected in the cortex. The entire field is therefore built up by a series of overlapping holographic patches.

The procedure is, curiously enough, something similar to what happens when mathematicians build up Einstein's curved space-time. In each region, space-time can be represented by a patch. Yet each patch cannot be continued smoothly onto its neighbor. The result is a patchwork quilt of space-times which together make up the curved geometry of general relativity. In a similar way, the visual field within the brain is made up of a series of overlapping holographs, each one representing part of the field.

Pribram points out that there are advantages in holographic patches as opposed to a single spread-out holograph. Such a patchwork system will be very sensitive to movement, for this will involve a difference between one patch and the next as the object moves across the visual field. Insects with compound or multiple visual fields arranged like the facets of a jewel are very sensitive to small movements around them, and the patch holographs may work in a similar fashion by conveying a three-dimensional moving image made up of overlapping holographs.

THE DOG AS A SINE WAVE

The next step in bringing the holographic analogy into brain science was to determine the actual nature of the transformation that takes place between world and brain. That is: How is the three-dimensional light-dark solid environment we move in translated into interference patterns moving along the optic nerve into the cortex, and, once there, how do we experience it as sight?

Ordinary holographic transformations involve a form of folding known mathematically as an "integral transform." An integral transform converts one shape to another by translating the first shape into a number of simpler forms.

A particular type of integral transform used in holography is the Fourier transform. It is an extremely efficient way to store and compare data and has even excited the interest of computer engineers as a new way of handling complex information. In performing a Fourier transform the intermediate step involves splitting up any complex shape, like the family dog, into the form of sine and cosine waves. Evidence suggests that this step is also performed in the brain. Information about the environment is thus converted into simple waves and these waves are transmitted and stored as interference patterns.

The breaking down of a complex wave form into its simplest elements, an essential step in a Fourier transform, can be illustrated by the sound of an orchestra. The French horn produces a note very close to a pure sine wave—the simplest and smoothest possible oscillation. The notes of the violin are sharp and jagged; the trumpet is smoother, but not as smooth as the French horn. Each instrument has a characteristic shape to its note, as well as a characteristic way the sound grows and decays.

The theory of Fourier transforms says that the note of any instrument, even of the whole orchestra itself, can be exactly duplicated by adding together simple sine waves. In an electronic synthesizer a collection of oscillators is coupled together, each producing a sine wave of a different frequency. To produce a horn note only a single sine wave is needed. To produce the sound of a violin or any other instrument the synthesizer must be fed with a series of sine and cosine numbers. These Fourier transforms indicate how much of each of the different sine waves must be added together. The result is an uncanny reproduction of an instrument.

The human hearing system works on a principle analogous

to the synthesizer, but reversed. Vibrations in the ear are picked up on the eardrum. The drum transfers these vibrations through an ingenious sytem of bone levers into a canal of fluid. Within this canal is an array of fine fibers or hairs. Each hair vibrates at a particular frequency.

When a horn sounds, its sine-wave vibration excites the fluid in the ear and causes only hairs sensitive to that frequency to vibrate. Each vibrating hair triggers nerve impulses which in turn tell the brain that a pure sine wave is sounding. If a violin's note is picked up, it will excite different series of hairs, each one or group responding to a single sine wave. In other words, the violin's note is split up into its elemental sine waves or Fourier components. The resulting message which reaches the brain is information on the intensities of each sine wave—in mathematical terms, a series of Fourier coefficients. Within the brain the impression of a violin is perceived.

The breaking up of a complex wave form into its Fourier components is perfectly general. In place of a sound wave one could substitute a wave breaking on the seashore, or a wave shape drawn on a piece of paper, or indeed a shape of anything. Fourier analysis can be applied to fluctuations of the stock market or the profile of the human face. Pribram believes that many of the brain's basic mechanisms are involved in translating perceptual experience, even daydreams, into Fourier transforms, many of which are distributed across the brain as consciousness and memory.

In 1968, Pribram received a note from Professor Fergus Campbell at Cambridge University which suggested good confirmation that the visual system performed a type of Fourier transform. Experiments on the visual system generally involve laboratory animals, in which the response pattern is monitored by probes within the brain. Earlier researchers had suggested that the visual system works by picking up pattern clues, for example edges and moving bars. Campbell had found, however, that when bar gratings were shown to the brain, it appeared to decompose them into sine and cosine waves, Fourier coefficients. In other words, cells in the visual cortex respond not to a pattern itself but to the Fourier components of that pattern!

Fourier transforms are also used in the mathematical representation of holography, and the combination of evidence for interference and Fourier decomposition of objects into frequencies indicates that something very similar in mathematical form to holography takes place in the brain. Pribram believes that in addition to vision and hearing, movement and physical action are also encoded in a Fourier form, in fact that everything we experience as perception and movement exists at another level as a Fourier "frequency domain" or implicate order.

How, for example, would you describe the movements of a man hammering or a woman jumping on a trampoline? Or could you tell anyone how to ride a bicycle? In the 1930s a Russian scientist, Nikolai Aleksandrovich Bernshtein, gave one answer to this problem. He dressed his subjects in black leotards and placed white disks on their joints, elbow, wrist, knee, and so on. The subjects were then set to perform an action and filmed against a black background. What showed up was a rather complex flowing movement in space recorded by the white disks.

Bernshtein discovered that this flowing movement could be Fourier-analyzed—that is, mathematically translated into sine waves or frequencies. The Fourier transform then allowed hammering or jumping to be understood *as a whole movement* (a collection of frequencies), rather than as a series of steps. This recalls Bohm's approach to movement as a whole action in itself rather than a series of discontinuous steps. Bernshtein's analysis was so successful that he could predict the next phase in a movement to a fraction of an inch. Unfortunately Bernshtein's work was not translated into English until 1967, so neurologists in the West learned of it quite late.

Pribram realized that just as the optical holograph enfolds the whole image in space, so a similar transform could enfold a whole movement, such as a dance. Bernshtein had showed how the whole of a movement could be described mathematically. Possibly the brain worked in exactly that way. That would mean we don't learn to ride a bicycle by compiling individual steps but by "tuning in" to the whole movement—getting arms, legs, and inner ear to respond to the same "frequency." The transfer of learning from one limb to the other may also take place through the transfer of a "whole action" or frequency.

An optical holograph is a static thing, a frozen instant in time. There are holographs which show movement, but this occurs in a rather superficial way, by placing together successive instants as in a moving picture. In suggesting that whole movements can be encoded as in the Bernshtein transforms, that time as well as space can be enfolded in the brain, Pribram clearly wishes to move beyond the confines of his original optical-holograph metaphor. Pribram and other neuroscientists speak persuasively for the holographic brain. Where other models portray the brain as an endless collection of bits, all mechanically (or electronically) interacting, this new model shows it as process-oriented, always doing things as wholes.

One significant problem remains unanswered. Granted memory is distributed holographically across the brain as a holograph is distributed across the photographic plate, but in holography, patterns are stored on the surface of the plate in arrangements of silver halide particles, similar to the way information in a computer is stored by arrangements of magnetic molecules on tape or disk.On what substance of the brain is the memory of how to ride a bicycle stored?

Pribram has suggested that interference patterns are stored across the membrane of nerve synapses (the gaps between nerve endings) as permanent changes in their electrical sensitivity. In optical holography, literally thousands of holographic images can be stored on top of one another on a single plate. Each of these images can be recalled separately by illuminating a scene or some aspect of a scene like the one originally stored. In an analogous way, Pribram proposes, the brain-cell synapses could contain thousands of holographic images. There are millions of such synapses in the brain. This model could account for the phenomenon of association—how one image or experience or idea recalls another somewhat like it, how perception leads to thought to perception to thought in the stream of consciousness, an unfolding and constant refolding of holographically stored memory.

It is known that even in periods when neurons aren't firing there is a constant slow flow of electrical activity taking place between synapses everywhere in the brain. Pribram says it is possible the holograms aren't stored as frozen images as they are on a plate but instead in this constant tide of electrical activity. This would make the brain's holograms process structures. A tone of voice, a smell, or the sight of a snowfall retrieves a pattern of memory embedded in the interference patterns of the brain's flowing electrical activity. A memory might be a kind of dissipative structure in this electrical flow: a dissipative structure made of interference patterns.

A fanciful picture of interfering wave fronts of electrical activity in the area of nerve synapses. Pribram believes that very subtle changes in electrical sensitivity in the area of the synapses store the holograms. A virtually infinite number of holograms could be stored together in these synapse areas all over the brain. The storage process may involve the constant flowing of electrical activity that takes place between synapses.

A wild new theory by Frank Barr, a research physician at the California Institute for the Study of Consciousness, proposes that the light-sensitive molecule melanin found throughout the body (including the skin where it causes suntan), may be a "holographic film" in the brain. Bohm has said that matter is a kind of condensed or "frozen light"—more accurately, light (or energy) moving at average speeds slower than the speed of light. Melanin has thus far proven strangely resistant to the usual chemical and physical analyses, so its precise structure remains unknown. But Barr says melanin, which is the most primitive universal pigment in living systems and which is involved in a huge number of biochemical interactions, directs the activities of other molecules and, in effect, "eats" light and converts it into other forms of energy in order to maintain and evolve matter. It is, he claims, a kind of slowed-down light molecule at the crossroads between biological matter and energy. In the brain, he believes, melanin acts as a "black hole" which makes possible the holographic patterns.

Pietsch is content to leave the question of storage material for the distributed interference patterns more vague, saying that the brain's holograms are stored in "phase space," which may be either in a number of different kinds of locations or brain processes, or in something nonphysical, perhaps akin to Sheldrake's morphogenetic fields. Obviously answering this question will ultimately be important to the fate of Pribram's model.

IN THE STRANGE LAND WHERE OBSERVER MEETS OBSERVED

There is a certain appealing irony in the fact that Karl Pribram began his scientific career as a behaviorist in association with the psychologist B. F. Skinner. Noted for his reductionist scientific approach, Skinner is the ultimate believer in a mechanical universe composed of interacting parts. Skinner's experimental approach has become known as the "black-box" theory of the brain.

In electrical engineering the concept of a black box is often used for theoretical discussions or experimental exercises. A student is presented with a sealed black box and the supervisor asks him what is inside. In this exercise the student isn't allowed to open the box but must deduce its contents by applying electrical signals at an input and monitoring the nature of the output. Inside the box may be an oscillator, amplifier, filter, or some more complicated electronic circuit. By correlating input with output the student should be able to deduce an exact circuit diagram together with the values of resistors, capacitors, and transistors within.

Skinner argued that this was the approach that should be applied to the brain. The brain is an enormously complex mechanism, and at our present level of knowledge the formulation of theories about it is a waste of time. But by measuring output against input, data on the brain's processes can be accumulated. The environment or the scientists' laboratory provides the input or stimulus. Behavior is the output or response. Hence Skinner's black-box approach to the brain means gathering information on stimulus and response.

The difficulty with behaviorism and the black-box approach is the conjecture that human behavior is "nothing but" stimulus and response. Obviously it also assumes that everything is mechanistic and that the observing scientists can remain totally separate from the box. Clearly, it's not a Schrödinger's cat box. To a strict behaviorist, all human activity is the result of the environment's action on the brain and its genetically determined programs. The effect of a stimulus can range from scratching the nose to painting a picture. Indeed, Skinner has suggested that a poet "has" a poem much as a hen has an egg. In his novel *Walden Two*, Skinner suggested a modern Utopia could result from the mechanical control of stimuli through what he called "social engineering."

While he was a young scientist, Pribram became embroiled with Skinner in a discussion about the theories of Wolfgang Köhler. Köhler, a leading gestalt psychologist, had suggested that the brain forms an exact image of the external world. If the eye sees a square object, then a square arrangement of cells is excited in the brain. Skinner asked Pribram what sort of image is formed when someone is mowing the lawn. The young man was forced to admit he had no idea. In that case, Skinner said, he'd continue to treat the brain as a black box.

Years later, armed with the holographic model, Pribram realized he could answer Skinner's question. Mowing the lawn is encoded as a complete movement, a total action in which external time and space relations are enfolded within the brain. Pribram knew Skinner could reply that Köhler said that the brain provides an *exact* image of the outside world. Is the outside world therefore enfolded space and time? Here Pribram came upon an unusual insight. If Köhler was to be taken literally, *then the transformations of the brain must be literal mirrors of the transformations of the external world. In other words, the world must be a holograph.*

At this point Pribram's son suggested that the neuroscientist should read the papers of David Bohm. After studying Bohm's works, Pribram realized that a fortuitous mirroring of insights had occurred. Just as Pribram's study of the brain suggested a holographic process, so Bohm's study of quantum theory suggested that the world outside, which the brain was observing and thinking about, is also holographic. In the instant Pribram read Bohm's theory, the observer met the observed. The search for the atom and the search for the engram had led to the same vision.

Or almost. Looking at the brain in terms of distributed patterns where images, sounds, and actions are encoded as wholes may be a great leap forward. However, it is important to note that in itself the explanation is mechanistic, just as earlier explanations of brain function have been. As Pribram points out, his research is devoted to showing the correlations between structures and processes of the brain and states of mind. But whether the electrochemical changes *are* consciousness (mind) or whether other dimensions are involved

Karl Pribram

lies far beyond Pribram's line of research. The discussions of vision given earlier provide an excellent illustration of the limitations of the holographic brain model on this question. If vision is thought of as the presentation of an image to the retina of the eye, then holographic process appears to give a good explanation. But vision is far more complex than this. It is an active, intentional process in which the eye seeks and explores. Although the whole scene may be taken in at a glance, only that portion which falls on the most sensitive part of the retina is fully seen in detail. In a series of rapid, jerky movements the eye traverses a visual scene over and over again, exploring features of interest, returning to boundaries, taking in visual clues. It is a process that is directed by the subject's search for meaning, his disposition, and his whole past history. Even looking at a static, flat object like a painting is an intentional activity unfolding in time.

The process of vision itself may be holographic or something like it, but what directs this process is mind, and the nature of mind is still an open question.

At this point, because Pribram's looking-glass brain requires a deeper understanding of matter and energy, it must shade off into the theories of Prigogine, Sheldrake, Bohm. For Prigogine and Jantsch the universe of interwoven autonomous dissipative structures cooperatively evolved the brain, and mind is somehow implicated in that *whole* process and is not strictly localizable in the brain. Sheldrake agrees that the mind is not in the brain and believes that individual consciousness is connected to the morphogenetic field of human consciousness lying beyond space-time; this field of the past guides and maintains the formation of our individual minds. With David Bohm's holographic universe we come full circle. For Bohm, mind, or consciousness, is enfolded over the whole of matter and such things as will and attention are ultimately movements of that whole. The movements of the whole are mirrored in each individual holographic brain. So brain is identical to mind—both are holographic. But mind is also infinitely more than its brain—for the same reason.

In discussing Bohm's theory, we asked: If the world is all flowing movement, why does it appear to us as separate things? In light of Karl Pribram's looking-glass model we can now put the question in more concrete terms: If the world is composed of frequencies and the brain is a frequency analyzer (itself made out of frequencies of matter), how does the three-dimensional solid world we know come into being?

The answer is as before: We have to learn it. We learn to respond mainly to certain frequencies and not to the constant transformations of frequencies. A few selected holograms become stabilized and apparently separate from one another into "things." The holograms, formed as memory, reinforce the impression of these separate things, and so the explicate space-time world we know evolves out of the implicate universe of waves and frequencies.

In its explicit form, life has survived by fleeing from predators and seeking its food. It has learned to deal with the explicit orders of things, to abstract certain clues. The brain as a complex frequency has emerged to take the "average" of frequencies, to analyze the world, search for clues, and separate objects and events from their general enviroment.

In the process it has also separated itself as the observer.

And so we open the box to peer at the cat—one flowing web of frequencies looking at others: quantum frequencies, cat frequencies, scientist frequencies. In our familiar universe these "things" all seem so complicated and confusing. But in the looking glass universe we see now, simply, how differently they are all the same.

"Now cut it up," says the Lion.

Universe
Without Edges

She returns to her place with the
empty dish.

At one point in her trip through the looking-glass, Alice finds the Jabberwocky poem. At first she can't read it at all. It seems to be in a foreign language. Then she realizes that since this is a looking-glass world she must hold the poem to a mirror. The poem she reads in the mirror, however, is still bizarre and unfamiliar. "Somehow it seems to fill my head with ideas only I don't exactly know what they are."

The languages of wholeness may be similarly puzzling. We sense what they mean and then the meaning eludes us. They are like vague forms emerging out of the mist—out of (as Prigogine might put it) the intense fluctuations of twentieth-century science.

It is tempting here to make a summary, a condensation or conclusive synthesis of the new theories that have been described in this book. But these theories are creative, living things and depict a universe which is itself creative. To draw conclusions about the theories would be to limit and play false to what are essentially explorations into the unknown—adventures that may lead to totally unexpected discoveries. They are theories like the White King's pencil, about which he complained: "It writes all manner of things I didn't intend."

Are these revolutions in theory harbingers of the paradigm revolutions described by Thomas Kuhn? Are we witnessing the rise of a looking-glass paradigm? When Kuhn wrote of revolutions he meant something that occurs only rarely in the evolution of science, not an everyday change in a theory or approach. The paradigm label has since come to be applied to almost any new idea. According to Kuhn's original meaning, the theory of Rupert Sheldrake could indeed represent a paradigm shift, and Pribram's holographic theory of mind might be the first step to a new exploration of consciousness. Prigogine's dissipative structures are firmly based within existing paradigms, yet in Jantsch's extensions the idea moves toward a new and powerful vision of order in a world of fluctuation and change. And what of Bohm? In a sense, his approach is not a paradigm shift at all, for what Bohm proposes lies beyond Kuhn's conception of scientific revolution. For Bohm, new directions, new ways of seeing, new paradigms, novel theories remain within the realm of the known. They are still part of a fragmentary approach that has confused the human race since the beginning of time.

Bohm's commitment to wholeness is extreme, but all of the

theories we have called looking-glass theories share it. That is their revelation and appeal. And if nothing else, they show that it is possible to leave the closed room of objectivity behind and think about wholeness as more than just a mystical affirmation. Rooted in the problems and concerns of their fields, the looking-glass scientists have demonstrated that, rationally and scientifically, we can learn to navigate in a universe without edges, yet still investigate what we perceive as its "edges," or phenomena, and participate in the formation of its laws. They have shown that it is possible to conceive of a universe without progress but with continued unfolding. And they have shown how wholeness may be infused in every part and particle of our lives. We will not endeavor to predict what acknowledging such a universe could mean for the whole conduct of human affairs, since it is possible such recognition could transform human consciousness itself.

Whatever fate the looking-glass theories may suffer in the vicissitudes of science, they certainly bring a conceptual transformation. For thousands of years wholeness has been mute. Now it can speak.

Who can tell what it will say?

Perhaps other theories will replace those we have explored here, theories which express wholeness more satisfactorily. Perhaps the fragmentary view will continue to dominate science. But the theories of wholeness are, at least, new expressions of an ancient insight and of a more ancient longing, one which will come now into dramatic conflict with the equally ancient longing to possess and control through knowledge and ownership the various separate things of this world, including ourselves.

If the past is any judge (and it may not be), the verdict of science on these theories will be harsh. We are already familiar with some of the standards that will be applied:

O Will the new theories receive confirming observations? (induction)

O Are the theories set out in a way that admits disproof? (falsification)

O Will the theories allow scientists to make accurate predictions about experiments? (induction, falsification)

O Will they support sufficient "puzzles" for further research? (paradigm)

○ Will they resolve problems unanswered by the theories now in vogue? (falsification, paradigm)

The answers to these questions will be made by scientists, whatever appeal the theories may have for the lay public.

Our excursions through the looking-glass are complete, but science's have only begun—if it wishes to stay there.

Niels Bohr once said that on Mondays, Wednesdays, and Fridays he tried to think up all the crazy ideas he could and then on Tuesdays, Thursdays, and Saturdays he would try to disprove them. Science doesn't just consist of prediction and calculation and experiments; it is the constant play of understanding the world. As children, most scientists were fascinated by the "big" questions of the universe. How did it all begin? What lies at the end of space? How do things move? What is everything made of? Science is a dialogue with the universe in which questions are asked and insights born. Understanding is a reaching out to nature, just as the baby explores its mother's face or favorite toy. This grasping at meaning, testing, ingesting, questioning, and formation of pictures in the imagination, map making and map shredding is an unceasing process for the scientist as for the child. The scientist's main task is to look for ways of continuing his dialogue with nature.

Science itself is a looking-glass. When we look into this glass we see wondrous things which may have a truth like that of Alice's strange worlds: a mixture of fantasy and metaphor, reality and something else, something not quite effable.

But the point has come in our discussion to abandon this image and return nakedly to the theories themselves—to recognize that the looking-glass has been another name for the ceaseless activity of science which in the end is nothing more than our own habitual and fanciful curiosity.

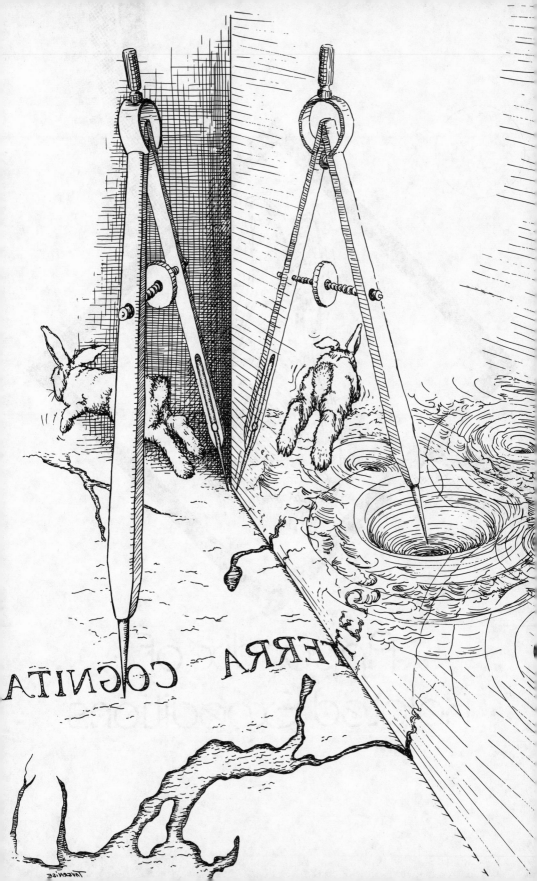

Appendix of
Related Expeditions

New scientific data suggesting the looking-glass spring up all the time in odd places. An IBM researcher reported using a mathematical technique called "fractals" which enabled him to analyze and computerize music very precisely. In one of his experiments, he constructed a computer model of a Stephen Foster song so that he could generate some songs that Foster might have written. The problem was, the computer kept coming up with songs Foster had actually written.

The looking-glass idea was not invented by twentieth-century scientists; they only discovered it in science. Renaissance painter and inventor Leonardo da Vinci wrote in his journal:

Every body placed in the luminous air spreads out in circles and fills the surrounding space with infinite likenesses of itself and appears all in all and all in every part.

Da Vinci also believed everything is the cause for everything else.

In *The Art of Fiction* Henry James described a novel as a web in which every point contains every other. Poets, musicians, and artists of all sorts have, through the ages, expressed the vision that the whole of the world is mirrored in each grain of sand or blade of grass or sequence of musical notes. Convincing evidence exists that creative works themselves are constructed on something like a holographic principle using the comparison-contrast dynamics of metaphor, irony, and other techniques as the patterning device.

Albert Rothenberg has studied creative leaps in artists' and scientists' work and is convinced creative process is based on what he calls "simultaneous opposites," or "Janusian thinking" (from Janus, the many-faced god). An example is Einstein's discovery of general relativity in the sudden image that a man falling off a roof would be simultaneously in motion and at rest. Another example is the simultaneous opposites in Mona Lisa's smile.

Douglas Hofstadter's recursion theory, entertainingly described in his Pulitzer Prize–winning *Gödel, Escher, Bach*, bears many resemblances to looking-glass theories. Hofstadter describes a reality or wholeness composed of continual returnings and reflections. From any of the many levels of this whole reality there is always something we can't see. At the center of the whole is an apparent incompleteness that might be compared to an eye at the center of a vortex. Artist Escher and composer Bach provide Hofstadter with illustrations of an order which can constantly turn back on itself like a Möbius strip. Mathematician Kurt Gödel provides a demonstration of how wholeness can include incompleteness. Armed with his idea of recursion, Hofstadter tries to show that the conflict between reductionism (the universe as parts) and holism (the universe as a whole) is really the same dragon biting its own tail.

There are obvious echos of the looking-glass in mystical traditions throughout the world. One of the most striking is *The Flower Garland Sutra*:

In the heaven of Indra, there is said to be a network of pearls, so arranged that if you look at one you see all the others reflected in it. In the same way each object in the world is not merely itself but involves every other object, and in fact *is* every other object.[53]

In the 1920s the Hindu mystical philosopher Aurobindo proposed that reality is composed of "vibrations" or frequencies, ranging from gross matter to subtle psychic energies (explicate to implicate). He predicted that this frequency view of reality would soon be discovered by science.

In his book *No Foreign Land: The Biography of a North American Indian*, Wilfred Pelletier describes how non-Western, "natural" cultures participate in a flowing looking-glass consciousness for the community as a whole. In this consciousness there are no observers, no "observeds." Action doesn't derive out of debate, political conflict, majority rule, or commands through a hierarchy—it comes out of a spontaneous emerging order.

Let's say the council hall in an Indian community needs a new roof . . . Well, everybody knows that. It's been leaking here and there for quite a while and it's getting worse. And people have been talking about it, saying, "I guess the old hall needs a new roof." So all of a sudden one morning here's a guy on the roof, tearing off the old shingles, and down on the ground there's several bundles of new, hand-split shakes—probably not enough to do the whole job . . . Then after a while another guy comes along and sees the first guy on the roof . . . Pretty soon he's back with a hammer or shingle hatchet and maybe some shingle nails or a couple of rolls of tarpaper. By afternoon there's a whole crew working on that roof . . . The whole community is involved and there's a lot of fun and laughter . . . All that because one guy decided to put a new roof on the hall. Now who was that guy? Was he a single isolated individual? Or was he the whole community? How can you tell?[62]

In the history of Western philosophy, looking-glass allusions abound; some have been mentioned. Contemporary philosopher Renée Weber of Rutgers, who has closely studied Bohm's work, finds parallels in it to Plato and Spinoza. Philosopher Frank McCluskey notes that the new science's emphasis on overcoming the subject-object distinction, ending the belief in absolute space, time, and motion, and the recognition that process is as important as product are all ideas contained forcefully in the philosophy of Hegel. Philosophers Liebniz, Heraclitus, Schopenhauer and Heidegger provide other entries into the looking-glass vision.

Perhaps the most articulate and penetrating modern philosopher for a looking-glass approach to reality is Jiddu Krishnamurti. For over sixty years, Krishnamurti has been arguing in countless lectures and books that "you are the world, the observer is the observed." Unlike the looking-glass scientists, he takes this insight directly into the realms of psychology, spirit, and the human condition. He offers no methodology for salvation, instead insisting that it is our search for methodologies that leads to fragmentation. He insists that if each individual will give attention to the fragmentations in his own consciousness, he will alter the consciousness of mankind. Among the specific fragmentations he frequently mentions are beliefs in nation-states, religions, and ideals. By giving reality and primacy to these fragments, he says, we create a world of violence and fear. Some of the looking-glass scientists have pointed out there has been no actual progress in science, though there has been change. Krishnamurti points out that we haven't made psychological progress either. We are not less violent than the savages; our violence is more refined, sometimes more subtle. He says the conviction that we can "better" ourselves individually or as a species through accumulating knowledge and skills only refines our fragmentation. An individual may consider himself morally advanced because he would not kill another human being, but to the extent that he participates in any form of prejudice, ambition, divisiveness—religious, political, personal—he contributes to the human consciousness which creates war and killing. Each of us is directly responsible for the whole of what humankind does. Fundamental change requires abandoning the "tribal," divisional thinking in *all* aspects of our lives. Krishnamurti has been a long-time friend of David Bohm.

It would be impossible to even mention all of the developments in which looking-glass vision or elements of it have cropped up independently or have been carried into other fields by thinkers inspired by the hard-science theories discussed in this book. New developments continue to unfold. Below a brief flavor only:

One of the most active areas for looking-glass ideas has been the psychotherapies. In the early 1970s, psychiatrist David Shainberg proposed what he called "the transforming process" in which patient and therapist are seen as mirrors of each other's consciousness. Shainberg viewed mental disturbances and pathology as rigid forms in consciousness. The patient is a patient because he fails to recognize that consciousness is continually in process. The goal of therapy is to assist awareness of this unfolding, which is the true identity or self. Erich Jantsch incorporated Shainberg's approach into his own synthesis of the dissipative-structure, co-evolution theory.

Psychiatrist Edgar Levenson has offered what he calls a "holographic model for psychoanalytic change," arguing that each moment in a therapy session or in consciousness reflects all others in the past. He advocates learning to "resonate" with these moments. Eugene Gendlin has developed a "focusing" therapy which aims at a "felt shift" in consciousness like the creation of a new dissipative structure.

Psychiatrist William Gray thinks that the unconscious mind is organized according to what he calls "feeling tones." *Brain/Mind Bulletin* reports that Gray made a study of criminals which suggests that they commit crimes which resonate with an unspoken unconscious personality pattern determined at birth. Treatment involves showing the criminal that there are more socially benign and productive activities which also resonate with one's personal pattern.

Recently there has been a veritable explosion of research into the relationship between the "frequencies" of thought and the "frequencies" of physiological processes like heart rate and the spread of cancer. Patterns of thought, it has been discovered, can directly alter even the patterns of "automatic" biological functions, and physiological behavior can alter thought. One psychologist teaches clients who have understood their psychological blocks but have been unable to act on this understanding to perform certain movements. When the body movement and posture corresponds to the understanding, he says, insight and actual change can occur.

Independently, in the mid-1970s, Ken Wilber, a consciousness researcher, developed a theory he calls "the spectrum of consciousness," which describes various levels of consciousness as parts of one continuum he calls "mind." Wilber has insisted that over the history of humankind these levels of consciousness have become fragmented and increasingly treated as if they were separate. He describes different Western therapies and Eastern religions as addressing different bands of this spectrum of consciousness, each speaking to what Bohm would call a relatively autonomous subtotality. He envisions ultimate freedom as the recognition that all the bands of the spectrum are really one whole movement of mind which extends beyond the individual into the universe at large.

In medicine, physician Larry Dossey has marshaled the looking-glass theories to argue that many diseases have their origin in a "time sickness" afflicting modern humans. He rejects the Cartesian notion that the body is a separate machine. He believes that by viewing the body as an ongoing process with the universe medical problems such as heart disease and cancer can be relieved and death (time) more harmoniously faced.

Marilyn Ferguson employs the theories of Prigogine, Bohm, and Pribram as evidence that human society is moving toward a new stage in evolution. A theory by Peter Russell announces the idea of a "planetary brain," in which human consciousness links up with nature.

In theology a new type of insight known as "process theology" has emerged, stimulated by the ideas of Alfred North Whitehead. Process theology proposes such ideas as that "God is adventurous" (He doesn't know what the result of his creation will be) and "each occasion is a selective incarnation of the whole past universe." The authors of this theory compare their new theology to St. Paul's insight that "we are members of one another."

Not only has Whitehead stimulated process theology, he has had direct influence on Bohm and the systems theorists, including Jantsch. He must be considered one of the founding fathers of a modern looking-glass approach.

In the study of the paranormal, Kenneth Ring has applied the hologram model to explain the experiences described by persons who have clinically died and been brought back to life. Ring says that the dazzling world of death they describe is Pribram's domain of pure frequencies: "The act of dying . . . involves a gradual *shift* of consciousness from the ordinary world of appearances to a holographic reality of pure frequencies."[63] Other researchers in the paranormal field have argued that since in a holographic universe time and space are collapsed, paranormal events such as precognition and telepathy become easily explicable. Noted parapsychologist Stanley Krippner believes that in light of the looking-glass theories, Jung's idea of synchronicity "will take on new meaning."

Presumably astrology could also claim a holographic connection, since it is well known that a good astrologer can read a chart from any number of angles. He can offer a reading by discussing the aspects all the planets make to each other, he can read it by concentrating only on one planet, or on simply the degree the planet takes in a sign. Each aspect of a chart reflects all others, though from a slightly different direction. The dramatic, if disputed, phenomenon of psychic surgery might also be explained by implicate dimensions. In psychic surgery the healer presumably materializes psychic energy imbalances and then pulls them out of the body and discards them as physical objects. Even the practice of voodoo, gaining power over someone by possessing some part of his body (hair, fingernails), might be given a holographic explanation.

The looking-glass perception is obviously rich, but also obviously risky. In thinking about wholeness perhaps we should remember what happened to Democritus' idea about atoms.

References and Further Readings

$$\left(\overline{\hspace{8cm}}\right)$$

Entries in this first list marked with asterisks are works wholly or in part accessible to the lay reader. Others require command of specific issues or scientific jargon. Works in the second list present no particular technical problem.

*1. Ardrey, Robert. *African Genesis*. New York: Dell, 1961.

*2. Bohm, David. *Causality and Chance in Modern Physics*. London: Routledge & Kegan Paul, 1957.

*3. ———. "The Enfolding-Unfolding Universe." Interview by Renée Weber in *Revision*, Summer/Fall 1978.

4. ———. *Quantum Theory*. Englewood Cliffs, N.J.: Prentice-Hall, 1951.

*5. ———. "The Physicist and the Mystic—Is a Dialogue Between Them Possible?" Interview by Renée Weber in *Revision*, Spring 1981.

6. ———. *The Special Theory of Relativity*. New York: W.A. Benjamin, 1965.

*7. ———. *Wholeness and the Implicate Order*. London: Routledge & Kegan Paul, 1980.

8. Brown, Robert. "A Brief Account of Microscopical Observations," *Philosophical Magazine*. Vol. 4, p. 161.

*9. Buckley, Paul and F. David Peat. *A Question of Physics: Conversations in Physics and Biology*. London: Routledge & Kegan Paul, 1974.

*10. Burr, Harold Saxton. *Blueprint for Immortality: The Electric Patterns of Life*. London: Neville Spearman, 1972.

11. Calder, Nigel. *The Key to the Universe*. London: Penguin Books, 1981.

*12. Capra, Fritjof. "Bootstrap Theory of Particles," *Revision*, Fall/Winter 1981.

*13. de Broglie, Louis, Louis Armand, Pierre-Henri Simon, et al. *Einstein*. New York: Peebles Press, 1979.

*14. Einstein, Albert. *Ideas and Opinions*. Sonja Bargmann, trans. New York: Crown Publishers, 1954.

*15. Globus, Gordon, Grover Maxwell, and Irwin Savodnik. *Consciousness and the Brain*. New York: Plenum, 1976.

*16. Goldsmith, Edward. "Superscience—Its Mythology and Legitimisation," *Ecologist*, Sept./Oct. 1981.

*17. Gombrich, E.H. *Art and Illusion*. Princeton: Princeton University Press, 1972.

 18. Gould, Stephen Jay. "Punctuated Equilibrium—a Different Way of Seeing," *New Scientist*, April 15, 1982.

*19. Heisenberg, Werner. *A Question of Physics*. New York: Harper Torchbooks.

*20. ———. *Physics and Beyond*. New York: Harper and Row, 1971.

*21. *The Holographic Paradigm and Other Paradoxes*. Ken Wilber, ed. Boulder, Colo.: Shambhala, 1982.

*22. "Interviews with David Bohm, Rupert Sheldrake and Renée Weber," *Revision*, Vol. 5, No. 2 (Fall 1982).

 23. Jantsch, Erich. *The Self-Organizing Universe: Scientific and Human Implications of the Emerging Paradigm of Evolution*. Oxford: Pergamon Press, 1980.

*24. Kuhn, Thomas. *The Essential Tension*. Chicago: University of Chicago Press, 1977.

*25. ———. *The Structure of Scientific Revolutions*. Chicago: University of Chicago Press, 1976.

*26. *Letters on Wave Mechanics*. New York: Philosophical Library, 1967.

*27. Lindsay, Robert Bruce, and Henry Margenau. *Foundations of Physics*. New York: John Wiley & Sons, 1936.

 28. Maturana, Humberto R., and Francisco J. Varela. "Autopoietic Systems: A Characterization of the Living Organization." Preprint from the University of Chile, Santiago, Chile.

*29. Pagels, Heinz. *The Cosmic Code: Quantum Physics as the Language of Nature*. New York: Bantam, 1982.

*30. Pietsch, Paul. *Shufflebrain: The Quest for the Hologramic Mind*. Boston: Houghton Mifflin, 1981.

*31. Planck, Max. *Scientific Autobiography and Other Papers*. Frank Gaynor, trans. Westport, Conn.: Greenwood, 1968.

 32. Prigogine, Ilya. *From Being to Becoming: Time and Complexity in the Physical Sciences*. San Francisco: W.H. Freeman and Co., 1980.

 33. Pribram, Karl, ed. *Central Processing of Sensory Input*. Cambridge: MIT Press, 1976.

 34. ———. "How Is It That Sensing So Much We Can Learn So Little?" In *The Neurosciences*, Karl Pribram, ed. Cambridge: MIT Press, 1974.

 35. ———. *Languages of the Brain*. Englewood Cliffs, N.J.: Prentice-Hall, 1971.

 36. ———. "Non-locality and Localization." Preprint from the Dept. of Psychology, Stanford University, Stanford, Calif.

37. ———. "Toward a Holonomic Theory of Perception." In *Gestaltheorie in der Modernen Psycologie*, S. Ertel, ed. Durnstadt: Steinkopff, 1975.
*38. Popper, Karl. *Conjectures and Refutations*. London: Routledge & Kegan Paul, 1969.
*39. ———. *The Logic of Scientific Discovery*. New York: Harper Torchbooks, 1959.
*40. ———. "Scientific Reduction and the Essential Incompleteness of All Science." In *Studies in the Philosophy of Biology*, F.J. Ayala and T. Dobzhansky, eds. Berkeley: University of California Press, 1974.
*41. Schilpp, A. E. *Albert Einstein: Philosopher-Scientist*. New York: Harper Torchbooks, 1959.
*42. Schrödinger, Erwin. *What Is Life?* and *Mind and Matter*. Cambridge, Eng.: Cambridge University Press, 1967.
*43. Sheldrake, Rupert. *A New Science of Life: The Hypothesis of Formative Causation*. Los Angeles: J.P. Tarcher, 1982,
*44. Thomas, Lewis. *The Lives of a Cell*. New York: Bantam, 1974.
45. Varela, Francisco J. *Principles of Biological Autonomy*. New York: North Holland, 1979.
*46. Watson, Lyall. *Lifetide: The Biology of Consciousness*. New York: Simon & Schuster, 1980.
*47. Weinberg, Steven. *The First Three Minutes*. New York: Bantam, 1977.
48. Wigner, Eugene. *Foundations of Quantum Mechanics: Proceedings of the International School of Physics, Enrico Fermi Course 49*. New York: Academy Press, 1971.

Two excellent sources of information on new developments in looking-glass science are:

Brain/Mind Bulletin, P.O. Box 42211, 4717 N. Figueroa St., Los Angeles, CA 90042.

Re-Vision, P.O. Box 316, Cambridge, MA 02138.

LOOKING-GLASS READINGS FROM OTHER FIELDS

49. Anderson, R. "A Holographic Model of Transpersonal Consciousness," *Journal of Transpersonal Psychology*, 1977, 9, pp. 119-128.
50. Aurobindo (Sri). *The Life Divine*. New York: Dutton, 1949.

51. Cobb, John V., Jr., and David Ray Griffin. *Process Theology: An Introductory Exposition*. Philadelphia: Westminster Press, 1976.

52. Dossey, Larry. *Space, Time and Medicine*. Boulder, Colo.: Shambhala, 1982.

53. Eliot, Charles. *Japanese Buddhism*. New York: Barnes & Noble, 1969.

54. Ferguson, Marilyn. *The Aquarian Conspiracy*. Los Angeles: J.P. Tarcher, 1980.

55. Gendlin, Eugene T. *Experiencing and the Creation of Meaning*. New York: The Free Press, 1962.

56. Hofstadter, Douglas R. *Gödel, Escher, Bach: An Eternal Golden Braid*. New York: Vintage, 1980.

57. Jung, C. K. "Synchronicity." Foreword to *The I Ching or Book of Changes*, Richard Wilhelm and Cary F. Baynes, trans. Princeton: Princeton University Press, 1967.

58. Krippner, Stanley, and J. White, eds. *The Future Science*. New York: Doubleday/Anchor, 1976.

59. Krishnamurti, J. *You Are the World*. New York: Harper & Row, 1972.

60. Levenson, Edgar A. "A Holographic Model of Psychoanalytic Change," *Contemporary Psychoanalysis*, Vol. 12, No. 1 (1975).

61. Monaco, Richard, and John Briggs. *The Logic of Poetry*. New York: McGraw-Hill, 1974.

62. Pelletier, Wilfred and Ted Poole. *No Foreign Land: The Biography of a North American Indian*. New York: Pantheon, 1973.

63. Ring, Kenneth. *Life at Death: A Scientific Investigation of the Near-Death Experience*. New York: Coward McCann, 1980.

64. Rothenberg, Albert. *The Emerging Goddess*. Chicago: University of Chicago Press, 1974.

65. Russell, Peter. *The Global Brain*. Los Angeles: J.P. Tarcher, 1983.

66. Shainberg, David. *The Transforming Self*. New York: Intercontinental Medical Book Corp., 1973.

67. Thomsen, Dietrick. "Making Music—Fractally," *Science News*, March 22, 1980, p. 187.

68. Whitehead, Alfred North. *Process and Reality*. New York: Macmillan, 1929.

69. Wilber, Ken. *The Spectrum of Consciousness*. Wheaton, Ill.: Quest, 1977.

Index

285

micro- vs. macroevolution in, 193–94
origin of life and, 195–99
co-homology, 138–39, 141
complementarity, 53
consciousness, 172, 234
of explicate vs. implicate order, 128–30
matter-mind continuum and, 127–28, 130–31, 133, 148
see also brain; memory; mind
cosmology, 132–34
creation theory, 185–86
creativity, 146, 154, 222–23
crystallization, 227

Dalton, John, 30–31, 33, 149, 229
Darwin, Charles, 161, 184, 185, 214, 217
deduction, 19
Democritus, 99
Descartes, René, 20, 21, 23, 24, 33, 127, 136, 149, 217
differential equations, linear vs. nonlinear, 173–75
dinosaurs, 235
dissipative structures (Prigogine paradigm), 126, 169–78, 211, 262, 270
as autopoietic systems, 178–80
critics of, 204–7
evolution and, 184, 191–92; *see also* co-evolution theory
implicate order theory combined with, 207–9
morphogenetic fields and, 234–35
in nonlinear realm, 173–75
order through fluctuation as dynamics of, 167–71, 237
paradox of, 169
Prigogine's thermodynamics studies and, 161–68
as self-organizing structures, 180–84
time irreversibility and, 172–73, 175–78
see also entropy
DNA, 185, 187–88, 195, 196, 214–15, 216, 218, 223
Donder, Théophile de, 161–62
double-slit experiment, 51–53, 89, 101, 107, 140, 231
Driesch, Hans, 218

Einstein, Albert, 35, 46, 49, 78, 86, 94, 106, 123, 127, 150, 217
atomic decay experiment of, 68–70, 82
hidden-variables argument of, 68–74, 88, 96, 97
observables as starting point for, 43, 44–45
quantum discovery of, 38–39, 41, 55–56
relativity theory of, 56–68, 81, 89, 101, 118, 258
unified field theory of, 67–68, 74, 75, 101, 144
Eldredge, Niles, 189–92
electricity, 138–39
electromagnetism, 75, 148
electrons, 75, 124, 165
in double-slit experiment, 51–53, 89, 101, 107, 140, 231
emission of, 68–70
locating of, 47
movement of, in metals, 76–77, 95–96, 163–64
observing movement of, 49–51, 53–54
orbits of, 35–40, 43
as "things" vs. "tendencies to exist," 54
as unfolding ensembles, 121–22
see also quantum mechanics; wave function; wave mechanics
elementary-particle paradigm, 27–28, 29
emission spectrums, 37–38, 40
empiricism, 21
encodings, 111–12, 115
energy:
conversions of, 155–56, 162; *see also* entropy
matter related to, 60,81, 132–33, 263–64
quantum fluctuations of, 80–81
engrams, 242, 245
ensembles, 120–22
entropy, 155–70, 175
on cosmic scale, 201–3
dissipating of, 169; *see also* dissipative structures
as increasing disorder, 156, 159–61
in living systems, 161–63
as molecular chaos, 159
in near- vs. far-from-equilibrium environments, 162–68
in open vs. closed systems, 162–63
order through fluctuation vs., 167–71
in reversible vs. irreversible processes, 155–56
time directionality and, 156–58, 172
epigenetic landscape, 188–89, 216–17
EPR, 71–73, 88, 97, 123, 124
equilibrium, 160,172
near-equilibrium vs. far-from-equilibrium states of, 162–68
punctuated, 190–92
ether, 56–58, 61
event horizons, 81–82
Everett, Hugh, III, 87–88, 102
evolution theory (neo-Darwinism), 161, 184–92
appearance of new species and, 232–33
competition axiom in, 184–85, 186, 192, 193, 199, 204
discontinuous jumps and, 189–92, 201

gene fluctuations and, 187–89
as hierarchical, 187, 200–201
morphogenesis and, 213, 214–16,
  221
objections to, 185–92, 215–16
see also co-evolution theory
explicate order, 126, 133
  experiencing of, 119, 128–30
  implicate order related to, 114–15,
    118–19, 124–25, 127, 131
  subtotals in, 119–21
eyes, see visual system

facts, making of, 105
falsification, 21–24, 29–30, 230, 253
Faraday, Michael, 138–39
Ferguson, Marilyn, 188, 206
fields:
  physical, 218–19, 220, 222
  see also morphogenetic fields
Finkelstein, David, 90
formative causation hypothesis, see
  morphogenetic fields
Fourier transforms, 259–62
Freud, Sigmund, 186

Gaia hypothesis, 198–99, 200, 234
Galileo, 63, 64
gamma rays, 50–51
Geller, Uri, 151
genes, 187–89, 196, 200
geodesics, 64–66, 68
geometry, 136, 137, 141
  pre-, 89–90, 136, 139, 142, 143
glycerine-dye experiment, 112–15,
  117–18
God, 127, 187
Gödel, Kurt, 23
Goethe, Wolfgang von, 154, 217
Gombrich, E. H., 26
Gould, Stephen Jay, 189–92
grand unification theory, 74–75,
  78–80, 82, 91, 102, 148–49, 151
Grassman, H. G., 141–42, 208
gravity, 75, 79–80, 148
  acceleration related to, 63–66
  as curvature of space-time, 64–67, 81
ground state, 40
guide waves, 140, 231–32

Hamilton, W. R., 141, 142
Hawking, Stephen, 21, 80–82
hearing system, 259–60
heat, as molecular motion, 158–59,
  160
Heidegger, Martin, 103
Heisenberg, Werner, 35, 41–45, 73, 90,
  97, 109, 217
  quantum mechanics of, 43–45, 46,
    47–49
  uncertainty principle of, 49–51,
    53–54, 55, 72, 80, 130–31
Heraclitus, 144, 161
hidden-variable theory, 68–74, 89, 90
  Bohm's approach to, 96–97, 139–40,
    146

EPR and, 71–73, 88, 97
wave function collapse and, 68–71
Hiley, Basil, 137–38, 139, 141, 142, 208
holographic model of brain function,
  109, 127, 128, 239, 245–68, 270
  experimental support for, 253–54
  Fourier transforms in, 259–61
  memory phenomena and, 245,
    252–55, 262–64
  movement and, 261–62
  storage mechanisms in, 262–64
  visual system and, 257–59, 260, 267
  world view and, 265–66, 267–68
holography, 109–11, 245–52
  as analogy for Bohm's holistic
    vision, 111–12, 115, 128, 184, 266,
    267
  interference patterns in, 111,
    245–49, 251
  laser light in, 249–51
  photography vs., 109, 111, 249
  recognition, 252
holomovement, see implicate order
  theory
holonomy, 120–21
Hume, David, 21

implicate order, 133
  experiencing of, 128, 130, 131
  explicate order related to, 114–15,
    118–19, 124–25, 127, 131
implicate order theory (Bohm's
  physics of holomovement), 93, 94,
  98–152, 196, 211, 237
  consciousness concept in, 127–31,
    133, 148
  cosmology of, 132–34
  dissipative-structure theory
    combined with, 207–9
  fragmentary vs. holistic approach
    in, 98–106, 205
  glycerine-dye analogy in, 112–15,
    117
  higher-dimensional realities in,
    122–26, 171, 199, 226
  hologram analogy in, 109–12, 115,
    128, 184, 266, 267
  language and mathematical
    expressions for, 134–39, 140–44,
    177n
  life principle in, 126–27, 167, 195
  morphogenetic fields and, 231–33,
    235
  movement concept in, 115–19, 126,
    261
  new approach to physics in, 120–22
  ocean current analogy of, 119–20
  order concept in, 106–8, 206
  paradoxes and dualities
    resolved by, 145–48
  relativity and quantum theories
    harmonized by, 100–103, 108,
    118–19, 148
  time concept in, 118, 125–26, 148,
    177, 234
induction, 18–21, 22, 24

inertia, 63n
insights, Bohm's concept of, 99–100, 131, 233
integral calculus, 115–16, 136
integral transforms, 259–62
interference patterns, information encoded in, 111–12, 115, 245–49, 251

Jammer, Max, 97
Jantsch, Erich, 161, 270
  on entropy at cosmic scale, 201–3
  self-organizing structures as concept of, 180–84
  see also co-evolution theory
Josephson, Brian, 87, 230

Kirlian fields, 218
Köhler, Wolfgang, 265
Kuhn, Thomas, 33–34, 76, 98, 105, 129, 185, 186
  on paradigm crises and shifts, 24–25, 28–29, 30–33, 73, 91, 149, 229, 230, 270

Lamarck, Chevalier de, 185
language:
  as explicate order, 119
  of science, 134–35
Laplace, Pierre de, 20, 21
laser light, 249–51
Lashley, Karl Spenser, 240–44, 252
Leibniz, Gottfried von, 115, 136, 139, 141, 150
L-fields, 218–19, 222
life:
  autocatalysis and, 167
  beginning of, 126–27, 167, 195–99
  entropy and, 161–63
  nonlife vs., 167, 220
light, 67
  as both wave and particle, 28, 38–39, 45
  speed of (c), 56–58, 118
linear differential equations, 173–75
linearity, 84–86
Locke, John, 21, 33
logical positivism, 21–22
Lorentz, Hendrik A., 47, 57–58, 61, 142, 177n
Lovelock, James, 198

Mach, Ernst, 21–22, 56, 149
magnetism, 138–39
Margulis, Lynn, 198
material frames, 142–43, 177n
mathematical approaches, 108, 136–39, 140–44
matter:
  energy related to, 60, 81, 132–33, 263–64
  -mind continuum, 127–28, 130–31, 133
  as only relatively stable, 144–45
matter waves, 45–46

Maturana, Humberto, 178–80
Maxwell, James Clerk, 38, 139
melanin, 263–64
memory:
  holographic model of, 245, 252–55, 262–64
  localized in brain, 240–42, 243, 244, 253
  long- vs. short-term, 242–43
  see also brain; consciousness; mind
Mercury, 67
mesons, 62, 78
metals, electron movement in, 76–77, 95–96, 163–64
microscope experiment, 49–51, 53–54
mind, 200, 239, 267
  -matter continuum, 127–28, 130–31, 133
  see also brain; consciousness; memory
molecular motion, 158–59, 160
morphic germs, 222–23
morphic resonance, 222
morphogenesis, 24, 211–35
  acquired physical traits and, 214, 225–26
  growth transformations and, 211–12, 223
  habitual movement and, 213–14, 224–26, 227
  neo-Darwinist views on, 213, 214–16, 221
  regeneration and, 213, 214, 219, 223
  regulation in, 212–13, 223, 243
  reproduction and, 213, 223
morphogenetic fields (Sheldrake's formative causation hypothesis), 218–35, 255, 264, 267
  adjustments and reinforcements in, 221–22, 232
  consciousness and, 233
  evidence of, 222, 226–31
  implicate order theory related to, 231–35
  in inanimate world, 220
  learned behavior and, 225–26, 227
  objections to, 230
  problems explained by, 223–26, 243
  transplants and, 223–24
motor fields, 224–25, 227
movement:
  absolute motion concept and, 56–58, 61–62, 63
  Fourier analyses of, 261–62
  habitual, 213–24, 224–26, 227
  in implicate order, 115–19, 126, 261
  Zeno's paradox of, 115–16, 118

natural laws, 19, 79, 171
  as indeterministic or probabilistic, 54–55, 68, 69–70
  as invariant, 20, 21, 24, 58–60, 63, 68–71, 74
  as protean, 33, 146, 171, 221–22
neo-Darwinism, *see* evolution theory

Neumann, John von, 73, 97
New Scientist, 97, 188, 230
Newton, Isaac, 20–21, 23, 24, 27–28,
    115, 136, 139, 150, 217
  absolute motion concept of, 56, 63n
  gravity theory of, 64, 67
Newtonian paradigm, 27–28, 56, 95,
    147
  shift from, 28, 35–43, 51, 60, 62, 67,
    79
  time-reversal symmetry in, 157–58,
    172
nonlinear equations, 173, 174–75
nonlocality, 88–90, 123, 146, 197
nuclei, disintegration of, 68–70

objectivity, 21, 33, 74, 128
  falsification standard of, 22, 24, 29,
    30
  in induction, 19–20
  in special theory of relativity, 58–60
observables, 43–45
observers, 237
  interaction of, with observed, 22, 33,
    49–51, 68, 130–31, 229
  moving at different speeds, 58–64
  as observed, 33, 34, 131, 148,
    171–72, 175
  observed separate from, 20, 21, 33,
    58–60, 68, 70, 130–31, 147–48
  Schrödinger's cat paradox and,
    82–88
  as unfolding ensembles, 122
order, 106–8, 146
  Cartesian, 107–8, 114, 115, 119,
    136–37, 139, 141, 142
  through fluctuation, *see* dissipative
    structures
  see also entropy; randomness

Pagels, Heinz, 89n
painting, schema for human form in,
    25–26
paradigms, 24–34
  practical value of, 26–27
paradigm shifts, 230
  choice of theory in, 28–32
  as perceptual changes, 24–25, 29,
    31–32
  scientific progress in, 31–32
paranormal phenomena, 151, 218
particle accelerators, 75, 77, 229
particles, subatomic, 75, 77–78, 79,
    105, 203
particle-wave duality, 28, 38–39, 45,
    122, 145
Pauli, Wolfgang, 41–42, 43, 95
Penfield, Wilder, 240, 242
Penrose, Roger, 89
perceptual mechanisms, 97–98, 99,
    129, 259–60
  see also visual system
photons, 50–52
  correlated pairs of, 88–89
Piaget, Jean, 119, 129, 137

Pietsch, Paul, 253–55, 256, 264
pilot-wave theory, 140
Pines, David, 95
Planck, Max, 38, 39, 41, 46
plants, holographic phenomenon in,
    255, 256
plasmons, 95–96, 163
Plato, 150, 221
Podolsky, Boris, 71
Popper, Karl Raimund, 21–24, 28–29,
    33, 45, 149, 253
pregeometry, 89–90, 136, 139, 142, 143
Pribram, Karl, 130, 244–45
  see also holographic model of brain
    function
Prigogine, Ilya, 154, 167–78
  nonhierarchical approach to reality
    of, 170–78, 187
  see also dissipative structures
protons, 35, 36, 75
punctuated equilibrium, 190–92

quanta, 38
quantum jumps, 39–40, 43, 49, 68,
    108, 115–16, 122, 136
quantum logic, 88
quantum mechanics, 43–45, 46, 47–49,
    54–55, 121–22, 125, 147, 165
quantum paradigm, 21, 23, 35–55,
    67–91, 132, 230
  Bohm's book on, 94–95
  crisis in, 74, 79–91
  Einstein's critique of, 68–74
  grand unification theory
    sought in, 74–75, 78–80, 82, 91,
    102, 148–49, 151
  mathematical approach
    appropriate to, 108, 136–39, 142
  in micro vs. macro world, 86–88,
    147
  observer's role in, 80, 82–88, 89n;
    see also uncertainty principle
  overlapping of relativity and, 79–82
  as paradoxical realm, 49–53, 82–88,
    90–91
  probabilistic universe in, 52–53, 68,
    69–70, 119
  puzzle solving in, 74–79
  relativity theory harmonized with,
    100–103, 108, 118–19, 148
  reversibility in, 172–73, 177–78
  rival theories in, 45–49
  shift to, 35–49, 73–74
  wholeness concept in (Copenhagen
    interpretation), 53–55, 62, 67, 68,
    72–73, 74, 86, 88–90, 97, 101, 105,
    127–28
quarks, 29, 77–78, 79

radioactive decay, 125–26
randomness, 107, 113, 146
  see also entropy
regeneration, 213, 214, 219, 223
Reich, Wilhelm, 255
Reimann, Georg Friedrich, 66